EW 102

A Second Course in Electronic Warfare

For a listing of recent titles in the *Artech House Radar Library*,
turn to the back of this book.

EW 102

A Second Course in Electronic Warfare

David L. Adamy

Artech House
Boston • London
www.artechhouse.com

Library of Congress Cataloging-in-Publication Data
Adamy, David.
EW 102: a second course in electronic warfare /David L. Adamy.
p. cm. — (Artech House radar library)
Includes bibliographical references and index.
ISBN 978-1-58053-686-7 (alk. paper)
1. Electronics in military engineering. I. Title. II. Series.
UG485.A3322 2004
623'.043—dc22 2004050666

British Library Cataloguing in Publication Data
Adamy, David
EW 102: a second course in electronic warfare. — (Artech House radar library)
1. Electronics in military engineering
I. Title
623'.043

ISBN 978-1-58053-686-7

Cover design by Igor Valdman

International Standard Book Number: 1-58053-686-7

10 9 8

Like EW 101 which came before it, this book is dedicated to my colleagues in the EW profession—in and out of uniform. Some of you have gone repeatedly into harm's way, and most of you have often worked long into the night to do things that are beyond the comprehension of the average person. Ours is a strange and challenging profession, but most of us can't imagine following any other.

Contents

Preface

EW 101 has been a popular column in the *Journal of Electronic Defense* (*JED*) for 10 years now—covering various aspects of electronic warfare (EW) in two-page bites on a month-to-month basis. The first 60 of those columns were reorganized into chapters with added material for continuity to become the book *EW 101*. The book, like the columns, has been very popular, but, since then, there have been almost 60 more *EW 101* columns. Some have provided deeper information on subjects covered in the first book and some are on entirely new EW areas. It was clearly time for another book—thus *EW 102*.

The target audiences for this book are the same as that for *EW 101*: new EW professionals, specialists in some part of EW, and specialists in technical areas peripheral to EW. Another target group is managers who used to be engineers—who now must make decisions based on input from others (who may or may not be trying to break the laws of physics). In general, the book is intended for those to whom a general overview, a grasp of the fundamentals, and the ability to make general-level calculations is valuable.

I sincerely hope that this book helps you be a better EW professional. The free world needs the best you have to offer in this important field of endeavor.

1

Introduction

This is the second book in the series, and like other books in that position it is intended to have stand-alone value—but not be redundant with the first book (in this case, *EW 101*). Like the first book, this one collects the material presented in many months of *EW 101* columns, organizes the material into chapters, and adds introductory and supplementary material for completeness—so it reads like a book rather than a collection of columns. It also has a feature requested by many readers of both the book and the columns: Solved problems.

This book comprises almost entirely new material that was not in the *EW 101* book. The exceptions are the sections which provide additional scope to material covered in *EW 101*. In those cases, there is a brief overview of the relevant *EW 101* material. The subjects covered in the rest of the book are:

- Chapter 2: Threats, from a functional and signal point of view. *EW 101* talked about threats in context, but never really focused in on them.
- Chapter 3: Radar Characteristics, is a functional discussion of the different kinds of radars—with emphasis on their significance to electronic warfare.
- Chapter 4: Infrared and Electro-Optical Considerations in Electronic Warfare, including heat seeking missiles, IR imagery systems, night vision devices, laser designators, and countermeasures.
- Chapter 5: EW Against Communications Signals, including radio propagation, digital communication, jamming, and various issues

regarding emitter location. Discussions are also extended to the EW impact of spectrum spreading.

- Chapter 6: Accuracy of Emitter Location Systems, including emitter location techniques (a brief review), error statistics, and circular error probable for all common emitter location techniques.
- Chapter 7: Communication Satellite Links, including the way satellite link performance is predicted, and link jamming.
- Appendix A: A large appendix of problems with solutions. These problems cover the subjects in both *EW 101* and *EW 102*, and the solutions are real solutions, with all of the important steps—not just answers.
- Appendix B: A cross-reference of the chapters of both the *EW 101* and *EW 102* books to the *EW 101* columns covering the same material.
- Appendix C: A list of reference books in electronic warfare and associated disciplines. Although not an exhaustive list of the books available on the subject, they are an excellent starting point for further, in-depth study of the field.

Like the first book in the series, and like the monthly tutorial articles from which both sprang, this book is intended to provide a top-level view of the broad, important and fascinating field of electronic warfare. Here are some generalities about the book:

- It is not intended for experts in the field, although it is hoped that experts in other fields and experts in subfields within electronic warfare may find it useful.
- It is intended to be easy to read. Technical material (contrary to popular opinion) does not need to be boring to be useful.

The coverage of technical material in this book is intended to be accurate as opposed to precise. The formulas, in most cases, are accurate to 1 dB, which is accurate enough for most system level design work. Even when much greater precision is required, almost all old-hand systems engineers run the basic equations to 1 dB first, then turn loose the computer experts to drive to the required precision. The problem with highly precise mathematics is that you can get lost down in the details and make mistakes of orders of magnitude. These mistakes are sometimes incorrect assumptions or (more often) an incorrectly stated problem. Order of magnitude errors get you (and probably your boss) into big trouble; they are worth avoiding.

When you work the problem to 1 dB, using the simple decibel form equations in this book, you will quickly derive a set of approximate answers. Then, you can sit back and ask yourself if the answers make sense. Compare the results to the results of other, similar problems...or just apply common sense. At this point, it is easy to revisit the assumptions or clarify the statement of the problem. Then, when you apply the considerable facilities, staff time, money, and (perhaps) stomach acid required to complete the detailed calculations they have an even chance of coming out right the first (or nearly the first) time.

1.1 Generalities About EW

Electronic warfare is defined as the art and science of preserving the use of the electromagnetic spectrum for friendly use while denying its use to the enemy. The electromagnetic spectrum is, of course, from dc to light (and beyond). Thus electronic warfare covers both the full radio frequency spectrum, the infrared spectrum, the optical spectrum, and the ultraviolet spectrum.

As shown in Figure 1.1, EW has classically been divided into:

- Electromagnetic support measures (ESM)—the receiving part of EW;
- Electromagnetic countermeasures (ECM)—jamming, chaff, and flares used to interfere with the operation of radars, military communication, and heat-seeking weapons;
- Electromagnetic counter-countermeasures (ECCM)—measures taken in the design or operation of radars or communication systems to counter the effects of ECM.

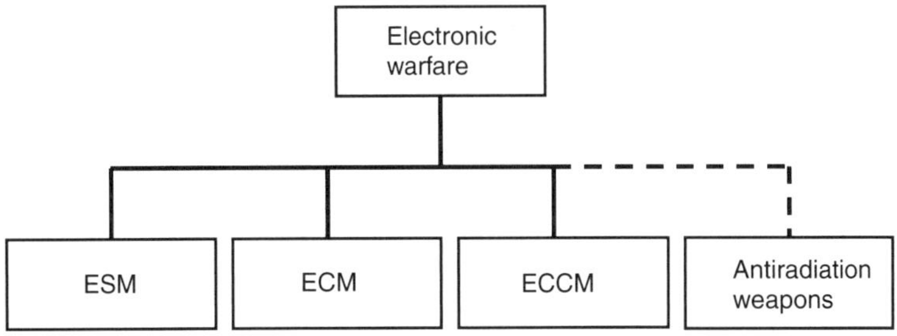

Figure 1.1 Electronic warfare has classically been divided into ESM, ECM, and ECCM. Antiradiation weapons were not part of EW.

Antiradiation weapons and directed energy weapons were not considered part of EW, even though it is well understood that they are closely allied with EW. They are differentiated as weapons.

In the last few years, the subdivisions of the EW field have been redefined as shown in Figure 1.2 in many (but not all) countries. Now the accepted definitions (in NATO) are:

- Electronic warfare support (ES)—which is the old ESM.
- Electronic attack (EA)—which includes the old ECM (jamming, chaff, and flares) but also includes antiradiation weapons and directed-energy weapons.
- Electronic protection (EP)—which is the old ECCM.

ESM (or ES) is differentiated from signal intelligence (SIGINT) [(which comprises communications intelligence (COMINT) and electronic intelligence (ELINT)], even though all of these fields involve the receiving of enemy transmissions. The differences, which are becoming increasingly vague as the complexity of signals increases, are in the purposes for which transmissions are received.

- COMINT receives enemy communications signals for the purpose of extracting intelligence from the information carried by those signals.
- ELINT receives enemy noncommunication signals for the purpose of determining the details of the enemy's electromagnetic systems so we

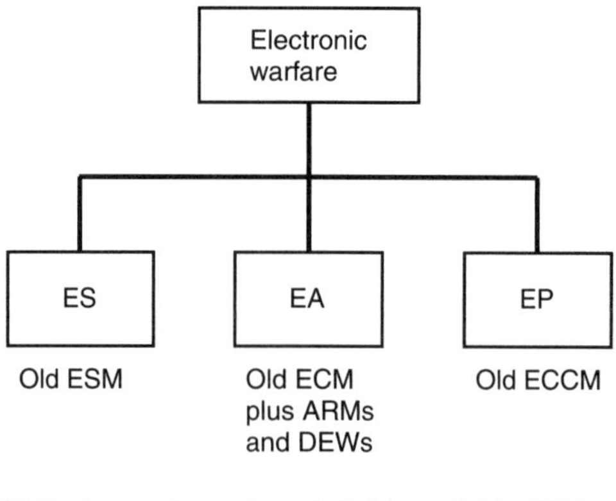

Figure 1.2 Current NATO electronic warfare definitions divide EW into ES, EA, and EP. EA now includes antiradiation and directed-energy weapons.

can develop countermeasures. Thus, ELINT systems normally collect lots of data over a long period of time to support detailed analysis.

- ESM/ES, on the other hand, collects enemy signals (either communication or noncommunication) with the object of immediately doing something about the signals or the weapons associated with those signals. The received signal might be jammed or its information handed off to a lethal response capability. The received signals can also be used for situation awareness, that is, identifying the types and location of the enemy's forces, weapons, or electronic capability. ESM/ES typically gathers lots of signal data to support less extensive processing with a high throughput rate. ESM/ES typically determines only *which* of the known emitter types is present and where they are located.

1.2 Information Warfare

A significant change that has occurred since the publication of *EW 101*, is the association of EW with information warfare (IW). EW is considered an integral part of information warfare—the action part. Information warfare includes actions taken to preserve the integrity of one's own information system from exploitation, corruption, or disruption, while at the same time exploiting, corrupting, or destroying an adversary's information system, as well as the process of achieving an information advantage in the application of force.

Figure 1.3 shows the so-called pillars of IW: psychological operations (PSYOPS), deception, electronic warfare, physical destruction, and operational security (OPSEC). These elements interfere with the enemy's ability to effectively use their military as shown in Figure 1.4. The OODA (observe, orient, decide, act) loop, shown in Figure 1.5 is the process required to take effective

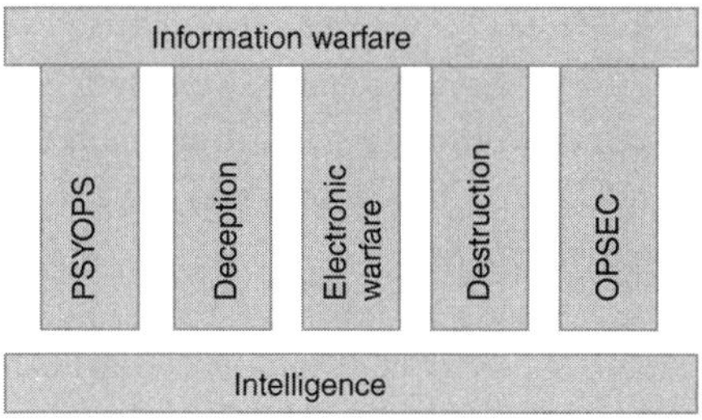

Figure 1.3 The pillars of information warfare are PSYOPS, deception, EW, destruction, and OPSEC, all supported by intelligence.

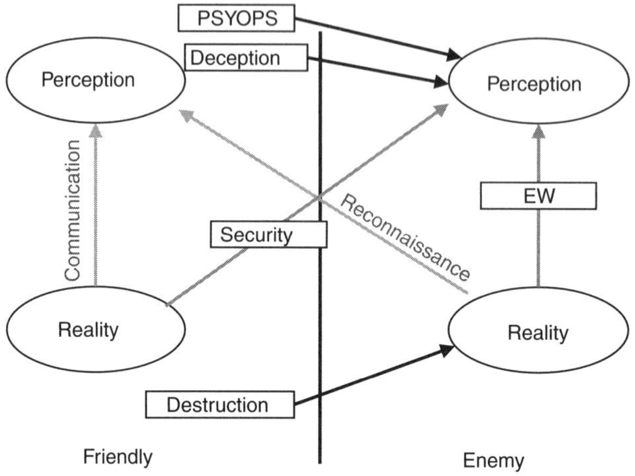

Figure 1.4 The elements of information warfare deal with friendly and enemy reality, and perception of reality.

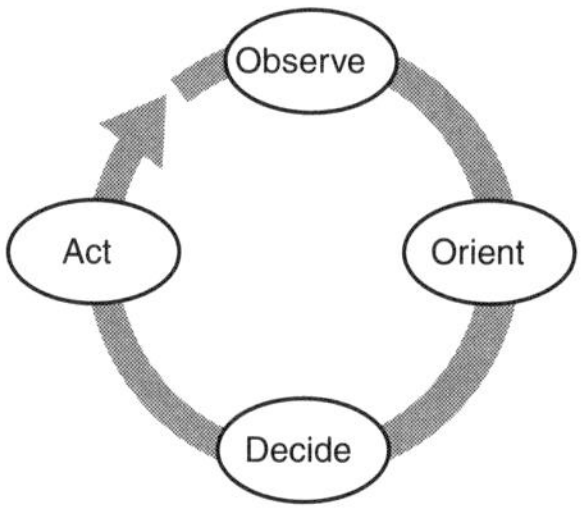

Figure 1.5 The OODA loop is the process involved in taking military action. IW interferes with the first three elements in the enemy's process.

military action. IW interferes with the first three steps in the OODA loop—with electronic warfare as the "action" element. This book focuses on EW, so we will not discuss IW further here, but it is important to understand the relationship between EW and IW in order to use EW techniques effectively in the current military environment.

1.3 How to Understand Electronic Warfare

It is the contention of the author that the key to understanding EW principles (particularly the RF part) is to have a really good understanding of radio propagation theory. If you understand how radio signals propagate, there is a logical progression to understanding how they are intercepted, jammed, or protected. Without that understanding, it seems (to the author) that it is almost impossible to really get your arms around EW.

Once you know a few simple formulas, like the one-way link equation and the radar range equation in their dB forms—you will most likely be able to run EW problems in your head (to 1-dB accuracy). If you get to that point, you can quickly cut to the chase when facing an EW problem. You can quickly and easily check to see if someone is trying to break the laws of physics. (OK, a piece of scratch paper is allowed as long as it is crudely torn from a pad—your colleagues will still class you as an EW expert when you get them out of trouble.)

2

Threats

Electronic warfare is, by its nature, reactive to threats. EW receivers are designed to detect, identify, and locate threats, and EW countermeasures are designed to reduce the effectiveness of those threats. In this chapter, we look at threats in general: the classes of threats, the platforms they threaten, the signals associated with them, and the classes of countermeasures used against them.

2.1 Some Definitions

Like most fields, electronic warfare is practiced by professionals who use their own special language. Unfortunately, this language is often a variance with proper usage in the native tongue of the land. To avoid confusion in later discussions, here are some important definitions associated with EW threats.

2.1.1 Threats Versus Threat Signals

Threats are the actual destructive devices and systems. In EW, we normally deal with the signals associated with the threat systems, so we often define a "threat" as a signal associated with the actual threat. While this can be confusing, it is the way people in our profession express themselves—a grammatical "sin" we have been committing for many years—and will continue to commit throughout this book.

2.1.2 Radars Versus Communication

We often divide threat signals into radar and communications classifications. The differentiation is that radar signals are used to measure location, distance, and velocity—while communication signals carry information from one point to another. While they have totally different functions, the two types of signals can have similar parameters. Radar signals can be either pulsed or continuous wave while communication signals are, by their nature, continuous (except for rare special cases). Radar signals are typically in the microwave frequency range, but can be as low as high VHF and extend up into the millimeter range. Communications signals can carry voice or data. They are typically considered to be in the HF, VHF or UHF frequency range. However, they can be found from VLF to millimeter wave. (There will be more about frequency ranges a little later.)

2.1.3 Types of Threats

Figure 2.1 is an overview of the types of threats to assets protected by various electronic warfare techniques. Note that this chart is somewhat controversial since some new threats cross the normal classification divisions. The purpose of this chart is to show the normally expected threat applications. As shown, radar-guided weapons are primary threats to aircraft and ships. The primary threats to ground mobile and fixed sites are weapons that home on laser-designated targets. Heat-seeking missiles are a primary threat to aircraft.

Lethal communication is described in Section 2.1.7. It is a primary threat to aircraft and fixed site assets—enabling a variety of types of weapons.

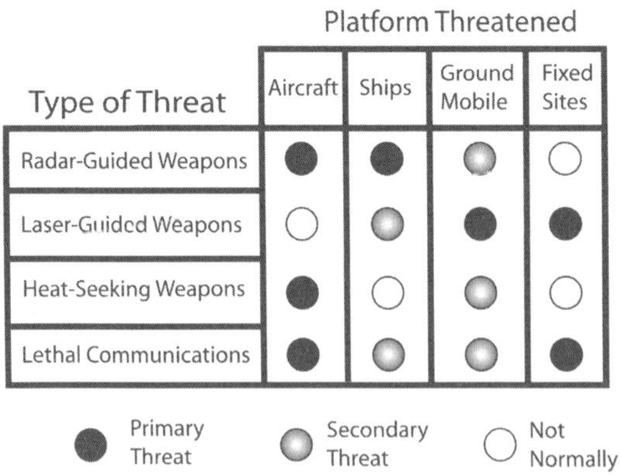

Figure 2.1 The various types of threats are normally threatening to these types of assets that are protected by EW systems.

2.1.4 Radar-Guided Weapons

As shown in Figure 2.2, radar is used to locate targets and to predict their paths of travel. A missile is guided to intercept the target. Note that the missile can be a rocket or a projectile (or many projectiles) from a radar-controlled gun. There are four basic guidance schemes that can be applied to radar-controlled weapons. Each has a different radar (or passive sensor) configuration and has strengths and weaknesses associated with the types of targets for which it is appropriate.

Ships are most commonly attacked by radar-controlled weapons. An aircraft (or other platform) locates a ship and identifies it as a target. Then, a missile is launched against the ship. Usually, the platform from which the missile is launched then leaves the engagement. When the missile gets close enough to the target to acquire it by radar, the missile homes on the ship, tracking its movement. The missile either attacks the target ship at the water line or makes a last minute vertical motion to strike it straight down through the deck.

2.1.5 Laser-Guided Weapons

Figure 2.3 shows an attack on a ground mobile target. The same technique can be used to attack a fixed asset, for example the pier of a bridge (the part most difficult to repair). In this type of attack, the laser must track the target so that a missile (which is typically fired by another platform) homes on the scintillation of the laser from the target. The designating platform can be either a manned or unmanned aircraft. It must remain within line of sight of the target during the whole attack.

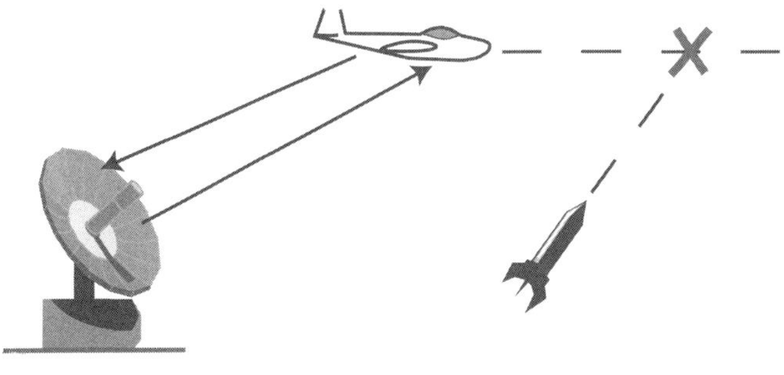

Figure 2.2 A radar-guided antiair threat determines the location and motion vector of a target aircraft to predict its flight path and cause a missile to intercept that flight path using one of several guidance approaches.

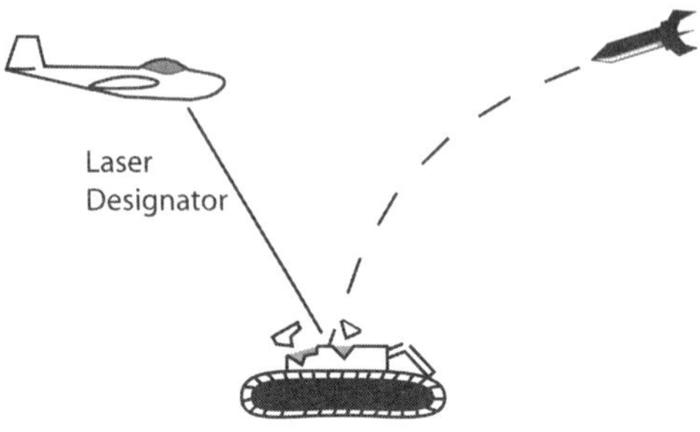

Figure 2.3 A laser-guided threat homes on the scintillation of a laser designator from a fixed or mobile target.

2.1.6 Infrared Energy: Guided Weapons

Everything emits some level of infrared energy (IR); the hotter the object, the more energy it emits. Since a jet aircraft engine is very hot, it provides a lucrative target for heat-seeking missiles. Early missiles attacked from behind the aircraft and homed on this high-heat target. Note that small, handheld weapons firing infrared missiles can be lethal to low-flying aircraft. IR missiles are used in air-to-air, ground-to-air, and air-to-ground attacks. Modern missile sensors can detect and home on the IR energy from targets at considerably lower temperature than that of a jet engine.

2.1.7 Lethal Communications

Lethal communications sounds like a contradiction in terms, since communication is merely the transfer of information. However, in almost all weapons above individual firearms, the information about the target's location and the ability to guide a weapon to the target are in different places. Thus, the sensor must transfer its information to some type of attack coordination center and that center must transfer acquisition and/or guidance commands to the actual weapon. The communication that transfers that information is extremely lethal.

Consider a simple example of lethal communication as shown in Figure 2.4. Artillery has killed more soldiers than any other type of weapon, and cannot typically engage targets without communication. The guns apply nonline-of-sight fire in response to calculated elevation, windage, and powder charge commands from a fire-control center. The fire-control center modifies its commands to the guns in response to communicated inputs from a forward observer who can see the target and the strikes of the rounds fired. Both communication paths are extremely lethal.

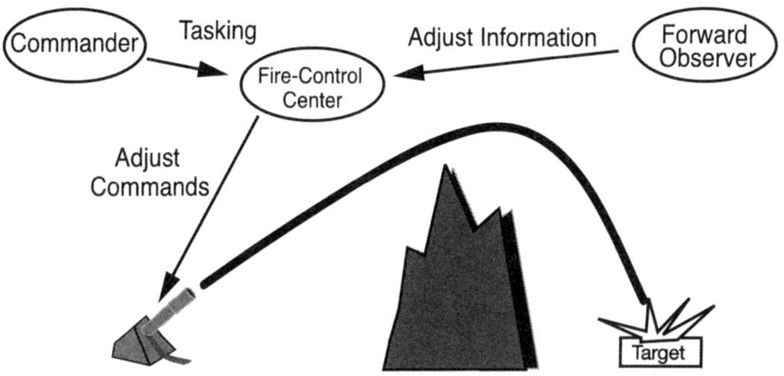

Figure 2.4 Artillery fire is adjusted to the target through lethal communication between a forward observer and a fire-control center and between the fire-control center and the guns.

2.1.8 Radar Resolution Cell

The resolution cell of a radar is the geometrical volume in which it cannot distinguish multiple targets. If there are multiple targets within the resolution cell, the radar will assume that only one target is present—at the weighted centroid of the individual target locations.

2.2 Frequency Ranges

Figure 2.5 shows the common names for frequency bands in the important threat range of 1 MHz to 100 GHz. This figure has three different columns, showing the three most common ways that frequency ranges are described. The left-hand column shows the common scientific notation. You will note that these bands divide at multiples of three. This is because each covers an order of magnitude of wavelength. For example, VHF is from 30 to 300 MHz—which corresponds to wavelengths from 1m to 10m.

The relationship of frequency to wavelength is given by the formula: $f\lambda = c$, where f is the frequency in hertz, λ is the wavelength in meters, and c is the speed of light (3×10^8 m/s).

The right-hand column shows the electronic warfare bands. The frequencies of threat radars are normally described in terms of these band designations. For example, D-band covers 1 to 2 GHz.

The middle column shows the official radar bands. Note that components (antennas, amplifiers, receivers, oscillators) are designated in catalogs in terms of these bands. It is also common to describe communications in terms of these bands. For example, satellite television broadcasting is done in C- or Ku-bands.

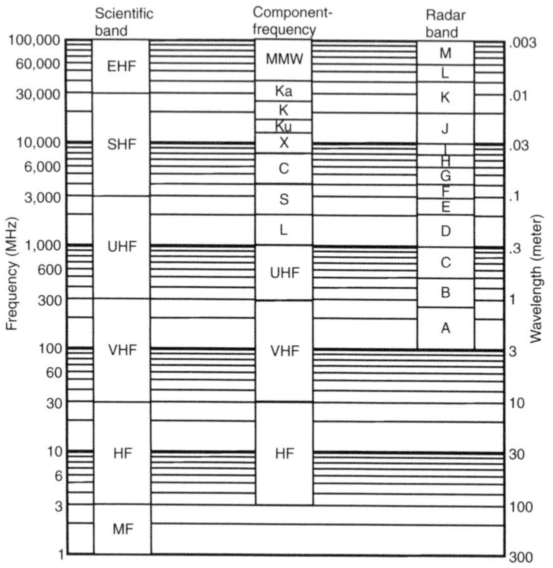

Figure 2.5 Frequency bands are designated three different ways: scientific band divisions, component-frequency ranges, and radar bands.

The frequency ranges of wideband receivers sold as HF, VHF, and UHF for EW and reconnaissance application vary from these frequency ranges: HF receivers often cover 1 to 20 or 30 MHz, VHF receivers often cover 20 to 250 MHz, and UHF receivers usually cover 250 to 1,000 MHz.

One very important point made by this figure is that it is easy to be confused by commonly used band designations. You will note that C-band can either be 500 to 1,000 MHz or 4 to 8 GHz. The best practice is to specify frequencies in megahertz or gigahertz when there is any confusion about band designations.

Table 2.1 describes the types of signal activity that take place in each frequency range. A generality about signal frequency is that transmissions become more dependent on line of sight as the frequency increases. HF and lower frequency signals can propagate clear around the world. VHF and UHF signals can propagate beyond line of sight, but are subject to severe nonline-of-sight attenuation. Microwave and millimeter wave signals are usually considered absolutely dependent on line of sight.

A second generality about frequency is that the amount of information carried by a signal is generally proportional to the transmitting frequency. This is because the amount of information carried depends on the signal bandwidth, and system complexity (antennas, amplifiers, receiver performance) is a function

Table 2.1
Typical Applications in Frequency Ranges

Frequency Range	Abbreviation	Type of Signals and Characteristics
Very low, low, and medium frequency (3 kHz to 3 MHz)	VLF, LF, MF	Very long-range communication (ships at sea). Ground waves circle the Earth. Commercial AM radio.
High frequency (3 to 30 MHz)	HF	Over the horizon communication, signals reflect from ionosphere.
Very high frequency (30 to 300 MHz)	VHF	Mobile communication, TV, and commercial FM radio. Severe losses beyond line of sight.
Ultra high frequency (300 MHz to 1 GHz)	UHF	Mobile communication, TV. Severe losses beyond line of sight.
Microwave (1 to 30 GHz)	μw	TV and telephone links, satellite communication, radar. Line of sight required.
Millimeter wave	MMW	Radars, data links. Requires line of sight. High absorption in rain and fog.

of the percentage bandwidth (i.e., the bandwidth divided by the transmitting frequency). Thus, signals carrying lots of information (e.g., wideband communication, television, or radar) are found at the higher frequencies.

2.3 Threat Guidance Approaches

There are four basic guidance approaches used by threat systems. These are active, semiactive, command, and passive. The type of guidance selected for a threat system depends on the nature of the platforms involved and of the typical engagement dynamics.

2.3.1 Active Guidance

Active guidance requires that a radar (or LADAR) be located on the weapon itself. Antiship missiles are an important application of this type of guidance. Once a missile is fired, it travels to the general area of the target ship, turns on its radar, acquires the ship, and guides itself to impact with the ship. Active guidance has the advantages that: the platform firing the missile can leave the area immediately after launch, guidance becomes more accurate as the range to the target diminishes, and it is very hard to jam at close range (because the radar power on the target is an inverse function of range).

2.3.2 Semiactive Guidance

In semiactive guidance (see Figure 2.6), the weapon has only a receiver. The transmitter is located remotely, for example, on the launching platform. The weapon then homes on the reflected signals from the target. When the guidance medium is radar, this is a bistatic radar implementation—very common in air-to-air missiles. Another important case of semiactive guidance is the laser-guided weapon that homes on the scintillation of a laser designator from a ground target. This type of guidance requires that the platform carrying the illuminator be present (and within line of sight of the target) throughout the engagement.

2.3.3 Command Guidance

In command guidance, a sensor—usually a radar—tracks a target to predict its path of travel. Based on the tracking information developed by the sensor, a weapon is guided to a point at which it will intercept the target (see Figure 2.7). The weapon has no information about the target location—it just goes where it is commanded. The classic example is a typical surface-to-air guided-missile system. One or more missiles are targeted and guided by a ground-based radar. Radar-controlled antiaircraft guns could also be considered to use command guidance because their projectiles are fired at appropriate azimuth and elevation angles and timed to explode at the aircraft's predicted position.

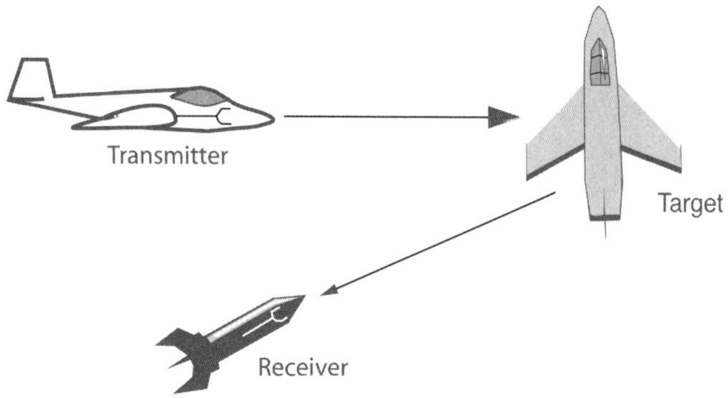

Figure 2.6 Semiactive guidance involves a receiver on the weapon and a remote transmitter. The weapon homes in on signals reflected by the target.

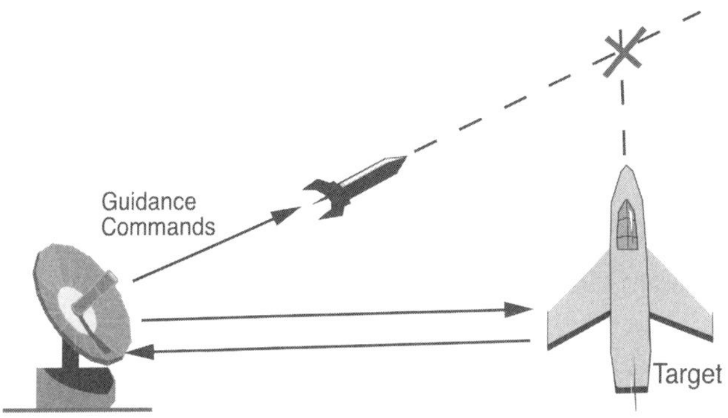

Figure 2.7 In command guidance, a radar remote from the weapon locates and tracks the target and guides the weapon to intercept with the target.

2.3.4 Passive Guidance

Passively guided weapons home on some emission from the target. Examples are: antiradiation missiles that home on radar transmissions and infrared missiles that home on heat radiated by targets (primarily aircraft). The weapon system does not radiate any type of targeting signal, so there is only one signal path—from the target to the weapon. Like active guidance, passive guidance allows "fire and forget" weapons. Thus, the launching platform (including individuals using shoulder-fired launchers) can leave the area or hide as soon as the weapon is launched.

2.4 Scan Characteristics of Threat Radars

Radars are designed to perform specific functions against specific types of targets under a particular set of conditions. In the practice of electronic warfare, it is useful to consider the way that the signals transmitted by a radar reflect the mission of the radar. We will consider acquisition radars and tracking radars in ground and airborne platforms, fusing radars, moving-target indicator radars, and synthetic aperture radars. We will consider the radar scan and the modulation on the transmitted signal, relate them to their appearance to an EW receiver, and then relate them to types of threat radars. Chapter 3 includes more detailed discussions of the signal characteristics themselves.

2.4.1 The Radar Scan

The radar scan (to an EW receiver) is the time history of the signal strength of the received signal. This is caused by the shape of the radar antenna beam and its angular movement relative to the location of the EW receiver. Figure 2.8 shows the antenna gain pattern of a radar antenna in polar coordinates (in one dimension). The antenna beam is shown as rotating (in that dimension) relative to an EW receiver location. Note that the main beam and side lobes all rotate past the EW receiver. Figure 2.9 shows the relative amplitude of the signal received by the EW receiver as a function of time. The shape of this curve can be analyzed to determine the beamwidth and scanning pattern of the radar.

2.4.2 Antenna Beamwidth

The radar typically has a narrow antenna beam which allows it to determine the azimuth and elevation of a target. The more accurately the radar must know the location of the target, the narrower the beam. The crossrange dimension of a radar's resolution cell is generally accepted to be the 3-dB antenna beamwidth. The radar can determine the actual angular location of a target within its resolution cell by small angular adjustments of the antenna pointing to peak the received signal strength. Figure 2.10 shows the time history of the received signal strength of the main beam of the antenna. If the rotation rate of the antenna is known, the beamwidth of the antenna can be derived from this figure. For example, if it is known that the antenna makes one full revolution in 5 seconds and the 3-dB beamwidth duration is 50 ms, the antenna beamwidth is 3.6°—determined from the following formula

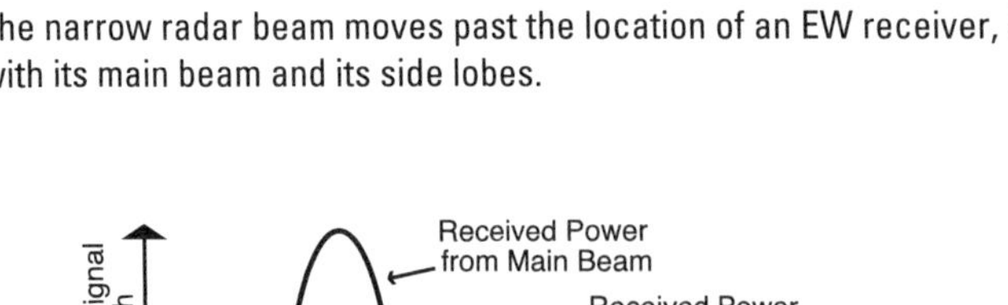

Figure 2.8 The narrow radar beam moves past the location of an EW receiver, illuminating it with its main beam and its side lobes.

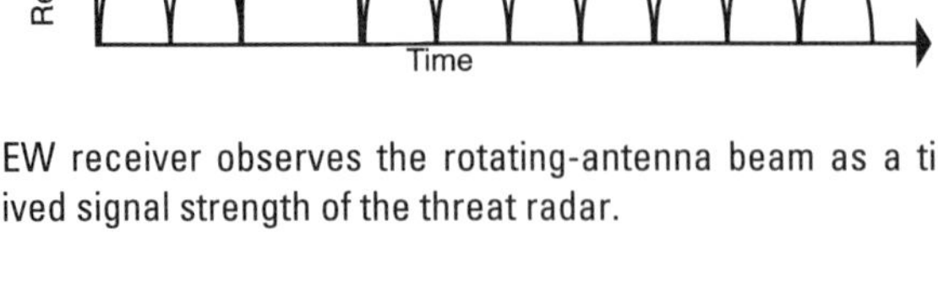

Figure 2.9 The EW receiver observes the rotating-antenna beam as a time history of the received signal strength of the threat radar.

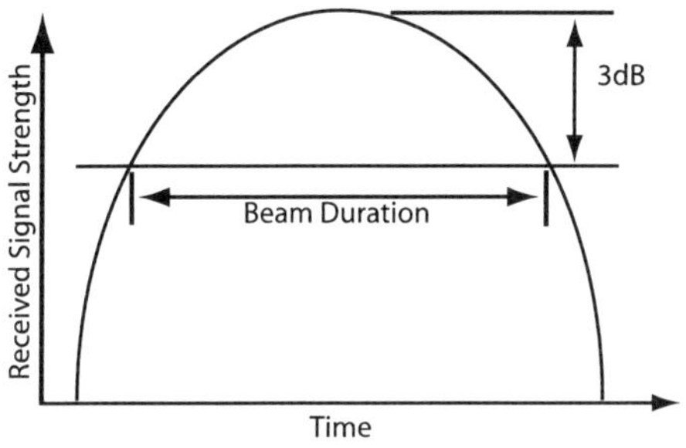

Figure 2.10 If the rotation rate of the antenna can be determined, the beamwidth can be derived from the time that the signal strength is less than 3 dB weaker than the peak receiver power level.

$$\text{Beamwidth} = \text{beam duration} \times 360°/\text{rotation period} = 50\text{ ms} \times 360°/5\text{ sec} = 3.6°$$

2.4.3 Antenna Beam Pointing

The pointing of the antenna is related to the operation the radar is performing. If the radar is trying to find a target, the beam will be swept over the angular area that might contain a target. If it is tracking a target it has already found, the beam on an older radar will be moved through a much smaller angular area around the target to enable the tracking function. In modern radars, which have multiple receiving sensors, the beam will get angular information from every pulse allowing it to follow the target (making it a "monopulse" radar). In the acquisition case, an EW receiver will see the antenna movement as received signal strength versus time. A monopulse radar beam is received as a constant level signal like that described next for a scan on receive-only radar. As examples of received radar beam patterns, let's consider: a ground-based search radar, a ground-based track-while-scan (TWS) radar, an airborne interceptor (AI) radar in search mode, a tracking radar with a conical scan, and a tracking radar with scan-on-receive only.

The ground-based search radar antenna will typically rotate 360° in azimuth (a circular scan). This causes an EW receiver to see evenly spaced main beams as shown in Figure 2.11. The time between main beams is equal to the period of rotation.

A ground-based TWS radar will typically cover an angular segment of many degrees. It tracks multiple targets in its angular range while continuing to look for more. For example, the SA-2 radar has two fan beams, one measuring the azimuth of every target in its field and the other measuring the elevation of every target in its field. The receiver sees a sector scanned beam as a power-time

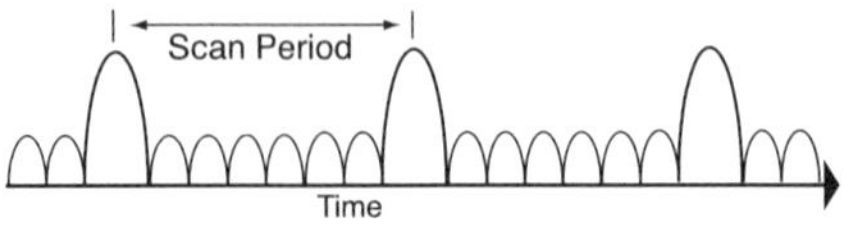

Figure 2.11 In a circular scan, the time interval between two receptions of the main beam is equal to the antenna-scan period.

history as shown in Figure 2.12 if it scans back and forth. If it "flies back" and scans its sector in one direction, the power history will show evenly spaced main beams, but will be identifiable as a sector scan from other situational cues.

A typical antenna for an airborne intercept radar in acquisition mode has a raster scan. This scan comprises a series of horizontal traces across a two-dimensional angular area—much like the way the beam in a television picture tube covers the face of the screen. As shown in Figure 2.13, the EW receiver will observe a scan pattern similar to the sector scan of the TWS radar, but the peak amplitude of the main beam will vary with the elevation of each scan relative to the receiver location. In this example, the radar beam passes through the EW receiver location on the second scan line.

A conical scanning radar uses a conical movement of its antenna beam to develop correction data to keep the target centered in its scan. As shown in Figure 2.14, the receiver does not see distinct main-beam patterns, but rather a sinusoidal variation of received power. The high point of the sine wave occurs when the antenna beam passes closest to the target—causing the antenna to rotate in that direction to center the target in the scan.

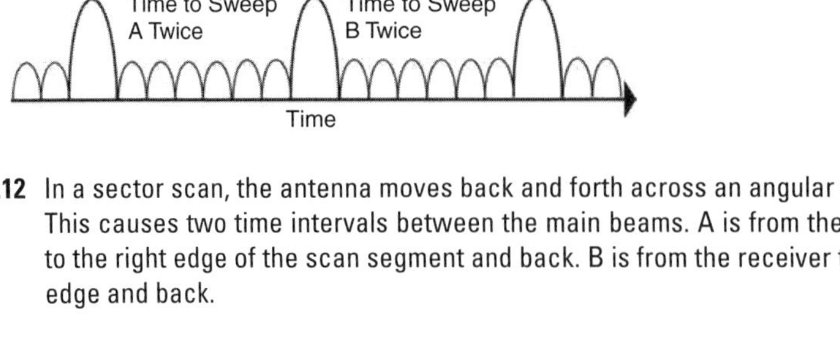

Figure 2.12 In a sector scan, the antenna moves back and forth across an angular segment. This causes two time intervals between the main beams. A is from the receiver to the right edge of the scan segment and back. B is from the receiver to the left edge and back.

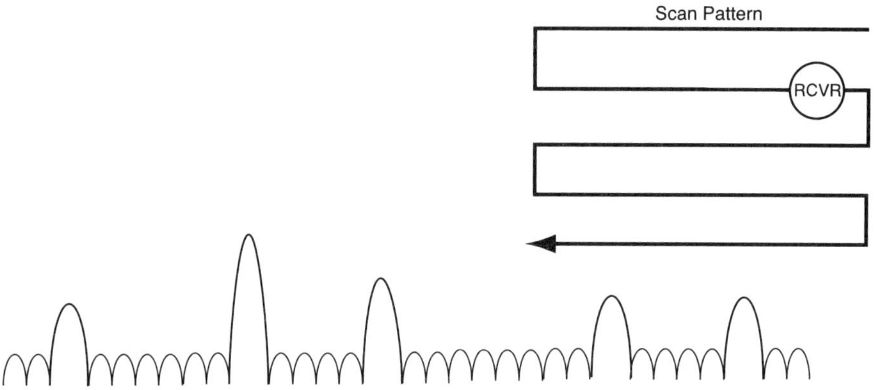

Figure 2.13 An EW receiver sees a raster scan as similar to a sector scan, but the main-beam peak amplitude varies with the distance to each of the horizontal scan lines from the receiver location.

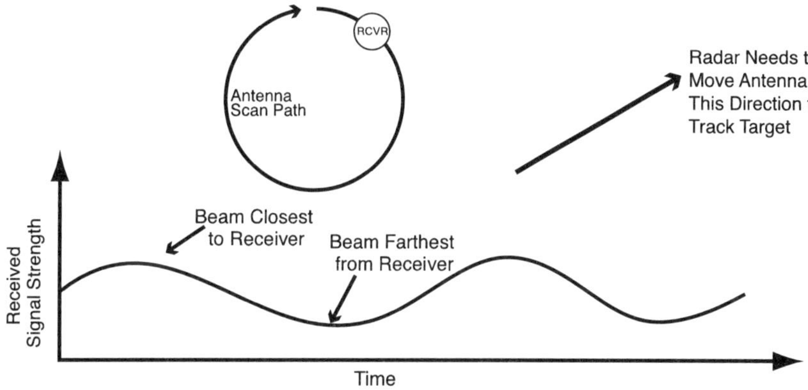

Figure 2.14 The receiver sees a conical scan as a sinusoidal variation in received signal power.

A scan-on-receive-only radar uses two antenna beams (typically generated by the same antenna). One is moved in a scanning pattern (for example, a conical scan) to receive return pulses from the target and calculate beam steering corrections. The second beam does not scan, but is pointed at the target using the correction information from the scanning receive beam. In this case, the EW receiver does not see any antenna scan, but rather the constant illumination of the transmit antenna.

2.5 Modulation Characteristics of Threat Radars

The modulation characteristics of a radar signal are dictated by the function of the radar. In this section we will consider pulse, pulse Doppler, and continuous wave radars in acquisition, guidance, and fusing applications.

2.5.1 Pulse Radars

Typical pulse radars output fixed frequency pulses separated by periods of silence during which echoes of those pulses are received. As shown in Figure 2.15, pulse modulation is characterized by the pulse width, pulse interval, and pulse amplitude. The pulse width (PW) is also called the pulse duration (PD). The pulse interval is the time from the leading edge of one pulse to the leading edge of the following pulse. The pulse interval in a signal is usually identified in terms of the pulse repetition frequency (PRF) or pulse repetition interval (PRI), but is also sometimes called the pulse repetition time (PRT). The pulse width and repetition rate are the same whether measured at the transmitter output, at the target, or at the receiver as long as the radar and the target are not moving, but the pulse amplitude changes a great deal.

The pulse amplitude in a radiated signal is the signal strength during the pulse. As the pulse leaves the transmitting antenna, this is the effective radiated power (ERP). When the pulse reaches a target, the pulse amplitude is the instantaneous power applied to the target. When the reflected signal reaches the radar receiver, it is the received signal strength.

The duty cycle of a radar is the ratio of the pulse width to the pulse interval. In regular pulse radars, this duty cycle can be from 0.1% to as much as 20%. This low-duty cycle means that the average power output by the radar is significantly less than its peak power. An important trend in radar development is the increasing capability of solid-state amplifiers to replace traveling wave tubes. This trend is moving duty cycles to 10% or greater in surface as well as airborne radars.

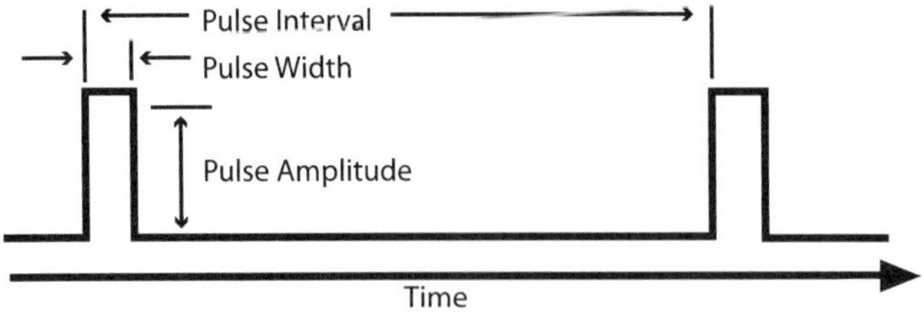

Figure 2.15 A typical pulsed radar modulation has a low duty factor with the pulse width being of the order or 0.1% of the pulse interval.

The maximum unambiguous range of a pulse radar is determined by the pulse interval. As shown in Figure 2.16, one pulse must have time to reach the target and the reflection must have time to return to the radar before the next pulse is transmitted. Otherwise, it would not be clear whether the received pulse is from the first pulse (reflected by a distant target) or from the second pulse (reflected by a much closer target). Since radar signals travel at the speed of light (3×10^8 m/s), the maximum unambiguous range is determined from the PRI by the formula:

$$R_{MAX} < 0.5\ PRI \times c$$

where

R_{MAX} is the maximum unambiguous range in meters;

PRI is the pulse repetition interval in seconds;

c is the speed of light in meters per second.

The minimum range of a radar is constrained by the pulse duration. As shown in Figure 2.17, the transmitted pulse must end before the leading edge of the pulse has time to propagate to the target and the reflection to propagate from the target back to the radar. A radar receiver typically blocks signals to its receiver while its transmitter is active, so the leading edge of the return from a longer pulse would be lost. The minimum range is derived from the pulse width by the following formula:

$$R_{MIN} > 0.5\ PW \times c$$

where

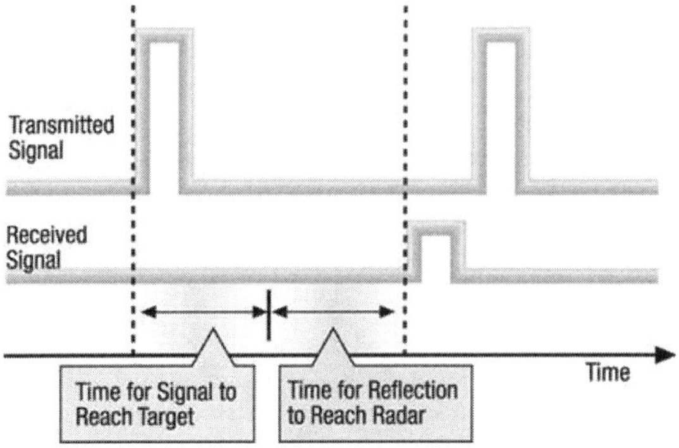

Figure 2.16 For unambiguous range measurement, the pulse interval must be more than twice the time for the signal to propagate from the radar to the target.

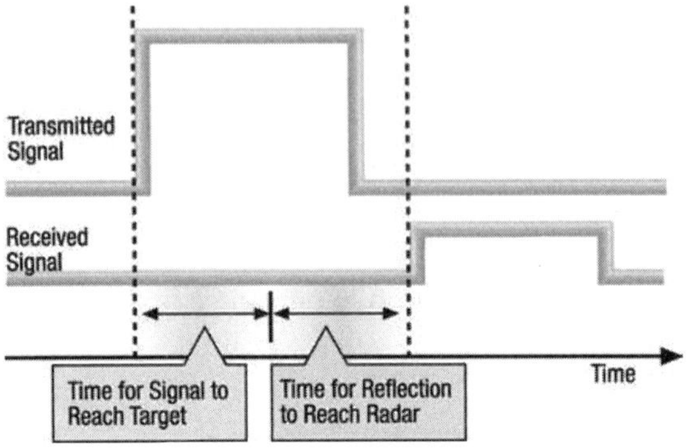

Figure 2.17 At the minimum operating range, the received pulse cannot begin before the transmitted pulse has ended.

R_{MIN} is the minimum range in meters;

PW is the pulse width in seconds;

c is the speed of light in meters per second.

For effective operation, a radar must place adequate energy onto a target. Since the signal strength of a transmitted signal decreases as a function of the square of the distance from the transmitter, a long range radar will typically have a long pulse width to increase the energy applied to the target as shown in Figure 2.18.

Because of these considerations, a short-range radar will tend to have short pulses and short pulse intervals, while long-range radars have long pulse width and interval.

The range resolution of a radar is determined by its pulse width. The longer the pulse width, the more crude the range resolution. Thus, long-range

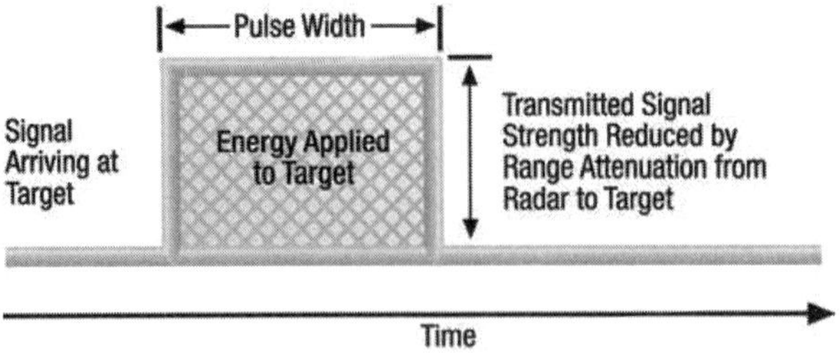

Figure 2.18 The energy applied to a target is the product of a pulse width and the signal strength at the target range.

radars with long pulse widths will have relatively poor range resolution. To improve resolution, "pulse compression" is enabled by adding either frequency modulation or digital modulation to transmitted pulses. When there is frequency modulation on the pulses, the radar is said to be "chirped," and the range resolution is improved by additional processing in the receiver. The digital modulation is binary phase shift keyed (BPSK) and allows the range resolution to be improved by a factor proportional to the number of digital bits during each pulse. Both of these pulse compression techniques are explained in Chapter 3. Note that a few modern tracking radars also use longer pulses with pulse compression.

2.5.2 Pulse Doppler Radars

Pulse Doppler (PD) radars are widely used in aircraft, and many ground-based radars also perform PD processing. PD radars use coherent signals. Coherent signal pulses are generated by transmitting intervals of a continuous reference signal. Since the signal is not continuous, a single antenna can be used (turning off the receiver during transmission), but the duty cycle is typically 10% to 40%. With this large duty cycle, the returning echo from a target can be lost because of the transmission of subsequent pulses. This occurs if the distance to the target makes the round-trip time equal to a multiple of the interpulse interval. The PD radar uses several pulse repetition frequencies (PRFs); each PRF causing a different pattern of distances to "blind ranges." In the PD processing, the returns from the PRFs which do not blank at the target range are considered by digital processing in the radar to determine the range to the target and the rate of change of range to the target. This range rate information allows the radar to distinguish signals returned by a target from signals returned by the ground for "look down, shoot down" operation. The rate of change of range to a target is the component of that target's relative velocity in the direction of the radar. The Doppler principle causes the frequency received by the radar to change by an amount proportional to the rate of change of range.

2.5.3 Continuous Wave Radars

Continuous wave (CW) radars use continuous signals rather than pulses, which means that they must have multiple antennas with adequate isolation to keep the transmitter from interfering with the receiver. They determine the range rate of targets from Doppler shift, and sometimes have frequency modulation which can be processed to determine range.

2.5.4 Threat Radar Applications

Threat radars are often classified as acquisition radars, tracking radars, and fusing radars. Table 2.2 shows the typical modulation parameters associated with each.

Acquisition radars search large areas to acquire targets. When targets are acquired, they are handed off to guidance radars. They are often called early warning/ground control intercept (EW/GCI) radars because they also provide target locations for controllers to guide fighter aircraft to their targets

Guidance radars are more directly associated with weapons. A guidance radar forms a track file on a target (i.e., a series of locations and velocities) so that a gun or missile can effectively attack the target.

The purpose of a fusing radar is to set off a warhead at the optimum distance from a target. For targets on the Earth's surface, this is typically a programmed distance above the ground. For airborne targets, the radar determines when the target is located within the burst pattern of the warhead—so the target will receive the maximum number of projectiles when the warhead explodes.

2.6 Communication Signal Threats

It is common practice in EW to call the signals associated with threats threat signals or simply "threats." As we have discussed, communication signals can be extremely "threatening," so the discussion of communication signal threats is very appropriate. Communication signals include both voice communication and digital data transmission.

Table 2.2
Range and Modulation of Threat Radars

Type of Threat	Operational Range	Modulation Parameters
Acquisition radar	Very long range.	Pulses, long PW, low PRF, often have pulse compression.
Tracking radar	Shorter range, lethal range of associated weapons.	Pulse, pulse Doppler, or CW. Short pulses, high PRF. Modern tracking radars can also have pulse compression.
Fusing radar	Very short range, a few times the burst radius of warhead.	CW or very high PRF pulses.

2.6.1 The Nature of Communication Signals

Communication signals carry information from one place to another, so they are one way by nature. However, most communication stations have transceivers (which both transmit and receive) allowing one-way propagation in either direction. This is important to communication intercept systems because only the transmitter can be located by an emitter location capability.

In general, communication signals have continuous modulations and tend to be very high duty cycle compared to radar signals. Historically, communication has taken place in the HF, VHF, and UHF frequency ranges using AM or FM modulation. However, with the increased use of unmanned aerial vehicles (UAVs) and communication satellites, microwave communication signals have become common. The wider the bandwidth of the signal, the more information it can carry per unit time. The higher the frequency of the signal, the more bandwidth it can have, but the more dependent the transmission path is upon line of sight.

In the following sections we will focus on two important types of communication signals as illustrative of the characteristics of communication threats. These are tactical communication signals and digital link signals.

2.6.2 Tactical Communication

Tactical communication signals include ground-to-ground communication, air-to-ground communication, and air-to-air communication. These signals are typically in the HF, VHF, and UHF bands and the transceivers have antennas with 360° azimuth coverage (see Figure 2.19). Whip antennas are the most common in ground-based communication stations, and folded dipole antennas are most common for airborne platforms. The use of nondirectional antennas allows communication to take place without knowledge of the location of the other end of the communication link. Since 360° antennas have low gain, it may be desirable to use direction antennas (such as log periodic) to communicate between fixed sites. These antennas provide more gain and isolation from undesired signals.

Tactical communication transmitters typically have from one to several watts effective radiated power, and the links operate over a few kilometers of range. Note that HF links can be much longer range (requiring more effective radiated power) because of the nonline-of-sight nature of HF propagation. Communication to and from aircraft in VHF and UHF also has extended range because of the greater line-of-sight distances. The information carried on tactical communication links can be voice or data, and voice can be carried in either digital or analog format. The information can be encrypted, and the signals can

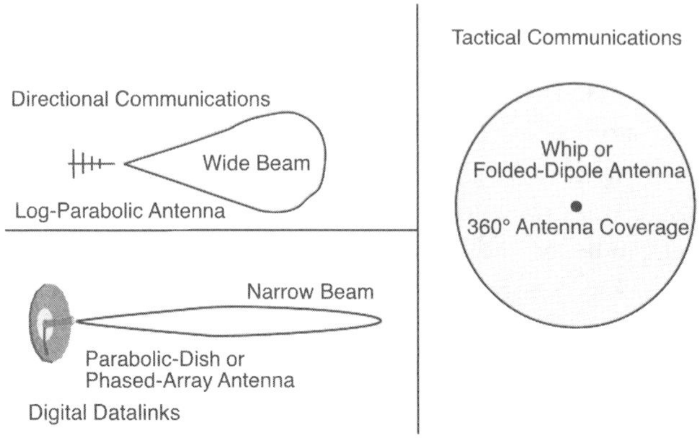

Figure 2.19 Antenna-beam coverage for communications signals depends on the application; 360° azimuth-coverage antennas are used when station locations are unknown. Directional antennas are used when station locations are known.

be either fixed frequency or protected from detection and jamming by use of any spread spectrum technique—frequency hopping being the most common.

Tactical communication often takes place in "push to talk" nets. This involves several transceivers operating at the same frequency, with only one station transmitting at a time. As shown in Figure 2.20, a typical net has one

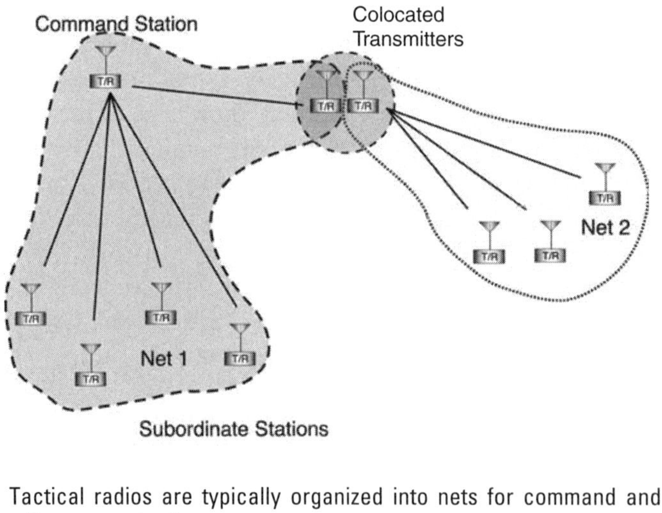

Figure 2.20 Tactical radios are typically organized into nets for command and control of military organizations.

command and several subordinate stations. Most communication takes place between command and subordinate stations, with the command station broadcasting at a significantly higher duty cycle than the subordinates. A single net (e.g., net 1 in Figure 2.20) is typically used by a single military organization. The nets of subordinate commands are interlocked with those of the higher command as shown in the figure. The two colocated stations (one in each net) indicate a lower level command location. The lower level commander uses a subordinate station in net 1 and the command station in net 2 (which will have a different frequency). One of the important reasons for the use of precision emitter location techniques is to identify these colocated stations.

Many tactical communication intercept systems include a display of frequency versus angle of arrival as shown in Figure 2.21. In a "battlefield situation," the signals will tend to be randomly spread in azimuth and frequency, with each dot on the screen representing one transmission by one transmitter. Subsequent transmissions by the same transmitter will show repeated hits at the same frequency and angle. One exception is the frequency-hopping signal which will have a series of frequencies at the same angle of arrival.

When this type of display is allowed to integrate for even a few seconds, it will be seen that virtually every frequency in a communication band is used. However, at any instant, there will be only 5% to 10% occupied channels. This is an important factor in the operation of tactical communication search systems.

2.6.3 Digital Data Links

Digital data links typically carry digital information at microwave frequencies. We will consider the UAV to control station links as typical examples. As shown in Figure 2.22, the UAV receives commands from a control station and returns payload data to that station. The command link (or "uplink") is usually

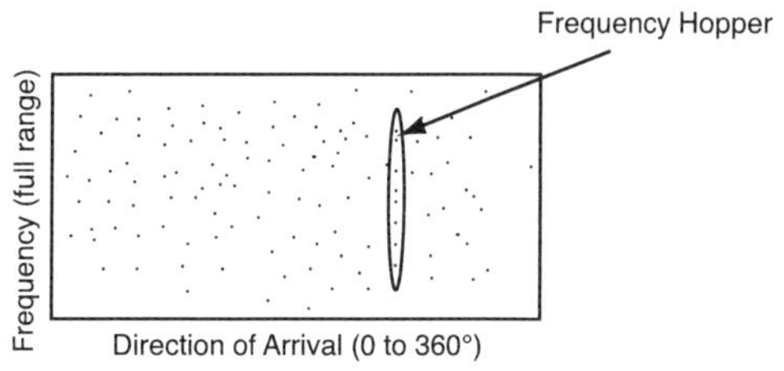

Figure 2.21 Tactical communications signals seen by a communications-intercept system are typically randomly spread over the frequency and angle of arrival. At any instant, 5% to 10% channel occupancy can be expected.

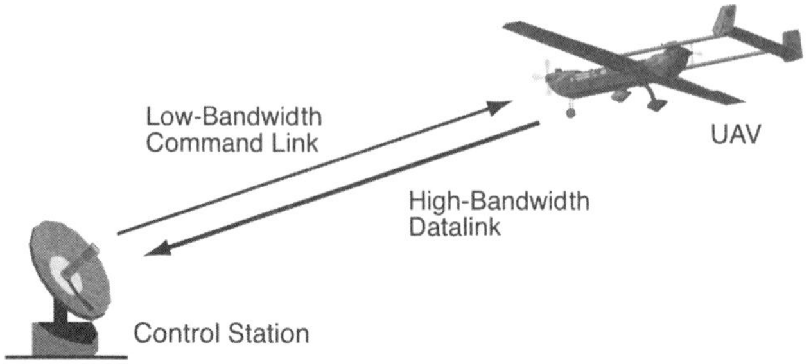

Figure 2.22 Links between UAVs and their control stations are typical digital datalinks.

narrowband, because command signals tend to have relatively low data rates. The uplink signals will typically be encrypted and have a high level of spectrum spreading. This protects the control station from detection and location by hostile emitter location systems, and makes it much harder for an enemy to interfere with control of the UAV or its payload.

The UAV to control station link is called the "downlink." It is also called the "data link" because it carries the output data from payloads. It typically has a much wider bandwidth that the uplink signal because it is carrying a great deal of information. The most common UAV payload is imagery (television or forward-looking infrared) which often requires millions of bits per second. These signals are typically encrypted and have some level of spread spectrum protection. However, the wide data bandwidth limits the amount of frequency spreading that can be applied.

Uplink antennas typically have narrow beamwidth, providing gain and making them harder for hostile emitter location systems to intercept. Downlink antennas are limited in size because of the size of the UAV airframe and aerodynamic considerations. Therefore, the downlink antennas typically have less gain and more beamwidth than the uplink antennas.

2.6.4 Satellite Links

Satellite links are important communication signals. They typically operate at microwave frequencies, carrying voice and data over great distances. Most satellites provide simultaneous access to many authorized users, so their signals are many megahertz wide. Some satellites support both commercial and military users. Typical commercial applications include television broadcast and telephone communication. Military satellites provide basically the same services, but the signal formats can be significantly more varied. Signals can be encrypted if appropriate, and spread spectrum can be employed for antijam protection.

3

Radar Characteristics

This chapter describes and discusses radar concepts and systems from the EW point of view. We will be looking at various types of radars to determine what they do, how they do it, and what their signals look like when viewed by an intercepting receiver. We will cover radar processing only to the level necessary to be able to discuss subjects like resolution, detection range, detectability, and vulnerability to jamming. Appendix C includes several recommended reference books that provide much more detail on radar theory and systems.

3.1 The Radar Function

The function of radar is to locate things by determining their range and angular position relative to the radar location and orientation. A radar determines the distance to some object (which we will call the target) by measuring the time it takes a propagated signal to travel to and from the target at the speed of light (see Figure 3.1). The range to the target is the speed of light multiplied by half of the time from the transmission of a signal to the reception of the same signal reflected from the target.

A radar determines the angular position of the target by means of a directional antenna with a gain pattern which varies as a function of angle from the antenna boresight. By comparing the amplitude of return signals as the antenna orientation changes relative to the target, the horizontal and/or vertical angle to the target from the radar's location can be computed. If the boresight of the radar antenna sweeps slowly through the angular location of the target, the

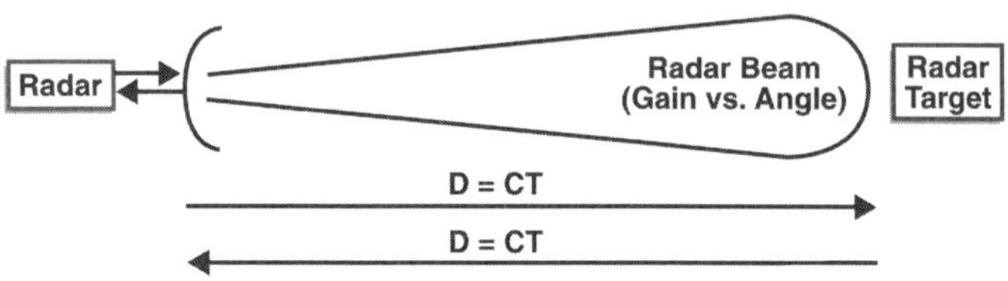

Figure 3.1 A radar determines the range to a target by measuring propagation time and its angular position by comparing the amplitude of return signals versus the orientation of the radar antenna.

target is determined to be at the angle(s) of the radar antenna when the peak received signal amplitude was measured.

Like most broad, sweeping statements, these are not always literally true. For example, some radars get additional angular information by processing return signals in the time domain. Still, although the process may be quite complex, there is always some underlying mechanism that traces back to those basic measurement functions.

Another characteristic of radars is that they look for consistency in the history of measured positions of objects they track. If a tracked object is moving relative to the radar, the radar processing will expect the tracked object to continue along the same path it has been following in the last few measurements.

3.1.1 Types of Radars

Radars are classified by modulation or application. The basic modulations separate radars into pulsed, CW, modulated CW, or pulse Doppler. Although there are many radar applications, the primary applications of interest to EW are target acquisition, target tracking, altitude measurement, mapping, moving target detection, and fusing.

There are also characteristics or attributes of radars by which we make further distinctions. Several of interest to EW considerations are:

- Radar implementations can be monostatic (with the transmitter and receiver at the same location) or bistatic (transmitting from one location and receiving return signals at another location).
- Monopulse radars measure target angle with information from each received pulse (rather than from sequences of several pulses).

- Track-while-scan radars can track one or more targets while continuing to look for more targets.
- Synthetic aperture radars use movement of the antenna along with sophisticated processing to create high-resolution radar maps.

3.1.2 Basic Radar Block Diagrams

As a convenient way to discuss some radar considerations important to EW, three basic radar block diagrams are considered.

Figure 3.2 is a basic (very basic) block diagram for a pulse radar. Pulse radars transmit short, high-power RF signals with a low duty cycle. Since pulses are being transmitted only a small percentage of the time, the same antenna can be used for both transmission and reception. The modulator generates pulses that cause the transmitter to output a high-power RF pulse. The duplexer passes the transmitted pulse to the antenna and the received reflected pulse to the receiver. Note that transmitted pulse is significantly higher in power than the received pulse, so there must be some provision to protect the receiver from reflected energy during the time the pulse is being transmitted. The receiver detects received pulses and passes them to the processor. The processor uses the amplitude of the received signal to perform tracking functions to keep the antenna pointed at the target if appropriate. It also performs range tracking to keep the radar focused on a single target. Information about the target location is output to displays. Control inputs include operating modes and target selection.

Figure 3.3 is a block diagram of a CW radar. It differs from the pulse radar in that its signal is present all of the time. This means that it must have two antennas, since it must receive a much weaker return signal while it is transmitting. The two antennas must provide sufficient isolation to keep the transmitted signal from saturating the receiver. The receiver compares the frequency of the received signal with the transmitted frequency to determine the Doppler shift caused by the relative velocity of the target. Modulation can be placed on the

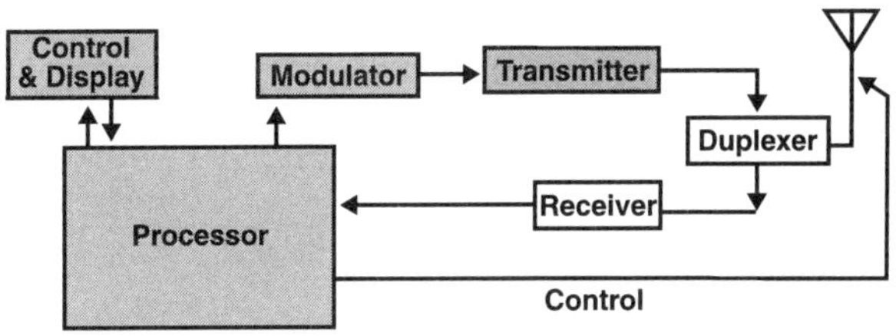

Figure 3.2 The transmitter and receiver of a pulsed radar share a single antenna.

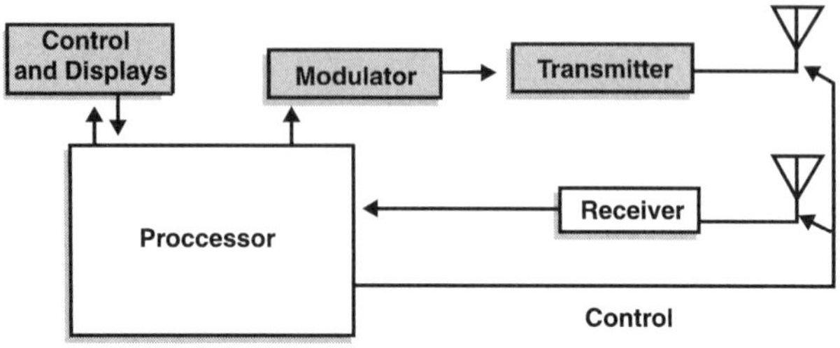

Figure 3.3 A CW radar must have separate transmit and receive antennas.

transmitter to allow the measurement of range. The processor performs target tracking and antenna control functions, and interfaces with the controls and displays as for the pulsed radar.

Figure 3.4 is a block diagram of a pulse Doppler radar. It differs from the pulsed radar in that the transmitted pulses are coherent. This means that the transmitted pulses are a continuation of the same signal and thus have phase consistency. Thus, the receiver can detect the return pulses coherently. As previously discussed for communications signals, coherent detection generally gives significant sensitivity advantages. It also allows the measurement of Doppler shifts so that the relative velocity of the target can be measured.

3.2 Radar Range Equation

The radar range equation is a widely used equation for the signal energy reaching a radar receiver as a function of transmitter output power, antenna gain, radar cross section, transmit frequency, time the radar illuminates the target, and range from the radar to the target. A common form of this equation is shown here.

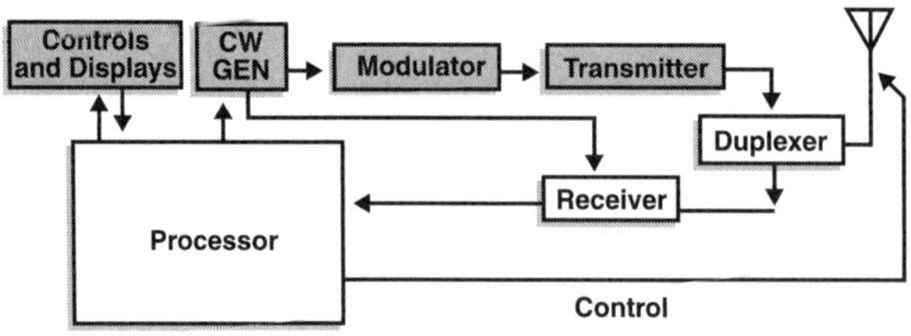

Figure 3.4 The pulse Doppler radar transmits a coherent signal and coherently processes return signals.

$$SE = \frac{P_{AVE} G^2 \sigma \lambda^2 T_{OT}}{(4\pi)^3 R^4}$$

where

SE = the received signal energy in watt seconds;

P_{AVE} = the average transmitter power (peak power × duty cycle) in watts;

G = the antenna gain (not in decibels);

λ = the wavelength of the transmitted signal (in meters);

σ = the radar cross section of the target (in square meters);

T_{OT} = the time the pulse illuminates the target;

R = the range to the target (in meters).

However, in electronic warfare, it is usually more useful to consider the received power in the radar receiver. It can be called the radar received power equation, and is often called (incorrectly) the radar range equation. This is the equation we use when calculating jamming-to-signal (J/S) ratios. Figure 3.5 shows the path of the radar signal and relates to this power equation. It has the implied assumption that the transmitter and receiver are colocated and have the same antenna gain. The most common form of the equation is shown here:

$$P_R = \frac{P_T G^2 \lambda^2 \sigma}{(4\pi)^3 R^4}$$

where

P_R = the received power (in any power units);

P_T = the transmitter power (in the same power units);

G = the antenna gain;

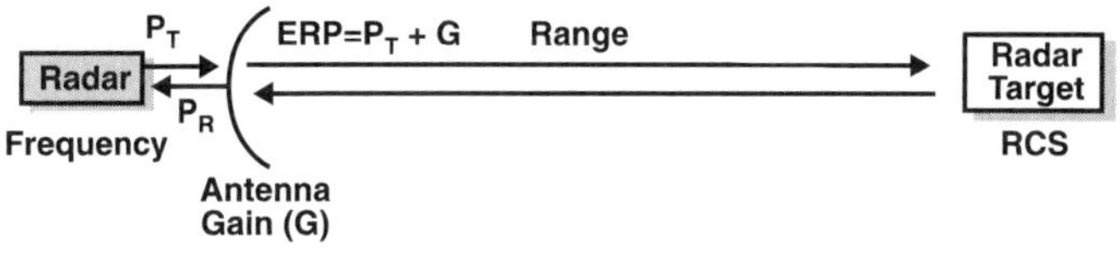

Figure 3.5 The radar received power equation determines the power into the radar receiver as a function of the transmitted power, the antenna gain, the transmitted frequency, the radar cross section of the target, and the range to the target.

λ = the wavelength of the transmitted signal (in meters);

σ = the radar cross section of the target (in square meters);

R = the range to the target (in meters).

A slightly different form of the equation uses the gain of the antenna in the transmit mode and the area of the antenna in the receiving mode (which makes the equation independent of frequency). In this form, the equation is written as:

$$P_R = \frac{P_T G A \sigma}{(4\pi)^2 R^4}$$

where A = the intercept area of the receiving antenna.

A form of the radar power equation frequently used in electronic warfare applications is derived by converting the first of the earlier equations into decibel form with the received power expressed in dBm, range in kilometers, and frequency in megahertz.

The wavelength term is converted to frequency by replacing λ with c /f (the speed of light divided by frequency). Then the constants and conversion factors are combined:

$$c^2 / \left[(4\pi)^3 (1{,}000\text{m / km})^4 (1{,}000{,}000 \text{ Hz / MHz})^2\right] = 4.5354 \times 10^{-11}$$

This value is converted into decibels, yielding –103.43 dB. The radar power equation then becomes:

$$P_R = -103 + P_T + 2G - 20 \log_{10}(F) - 40 \log_{10}(D) + 10 \log_{10}(\sigma)$$

where

P_R = received power in dBm;

P_T = transmitter output power in dBm;

G = antenna gain in decibels;

F = transmit frequency in megahertz;

D = radar to target distance in kilometers;

σ = radar cross section in square meters.

You will note that the constant in the decibel form of the radar power equation is reduced to 103. This is appropriate if the calculation need not be more precise than 1 dB. Otherwise, the 103.43 constant can be used. It is also common usage to shorten $\log_{10}(X)$ to $\log(X)$ in decibel equations. Please note that this type of decibel equation is only correct if exactly the proper units are used. If any other units are used (for example, nautical miles versus kilometers for range), the constant will need to be modified.

3.2.1 Radar Cross Section

The radar cross section of the target (RCS) usually represented by the symbol σ, is a function of its geometric cross section, reflectivity, and directivity.

- Geometric cross section is the size of the target as viewed from the aspect of the radar.
- Reflectivity is the ratio of the power leaving the target versus the radar power which illuminates the target. The rest of the power is absorbed.
- Directivity is the ratio of the power scattered back in the direction of the radar versus the amount of power that would have been reflected to the radar if the total reflected power were scattered in all directions.

The formula for RCS is:

$$\sigma = \text{Geometric cross section} \times \text{Reflectivity} \times \text{Directivity}$$

The RCS of an actual target, for example, an aircraft or a ship, is a vector sum of the reflections from each part of the physical object. It is typically very irregular with aspect angle and varies with the frequency of the radar.

The radar cross section of a target can either be measured in an RCS chamber or determined from a computer simulation. The RCS chamber is a specially instrumented anechoic chamber which measures the radar returns from actual targets, parts of targets, or scale models of targets. Computer RCS models are developed by representing the target by a number of reflecting surfaces (cylinders, plates, and so forth) and calculating the overall RCS from the phase adjusted combination of reflections from all of these surfaces.

As shown in Figure 3.6, there is an effective "gain" in the radar signal transmission path that is a function of the RCS. The equation for this gain is the following:

$$G = -39 + 20\log(F) + 10\log(\sigma)$$

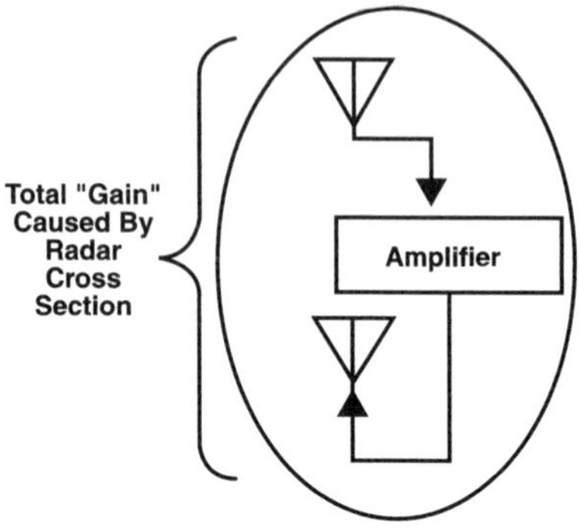

Figure 3.6 The radar cross section creates an effective signal "gain" that is functionally like the sum of the gains of two antennas and an amplifier.

where

G = the ratio of the signal leaving the target to the signal arriving at the target (both referenced to isotropic antennas) (in decibels);

F = the transmitted frequency (in megahertz);

σ = the radar cross section of the target (in square meters).

Figure 3.7 shows the RCS versus angle in the yaw plane of a typical aircraft, and Figure 3.8 shows the RCS of a typical ship from about 45° elevation as a function of horizontal plane angle from the bow. The units of the charts are dBsm [i.e., decibels relative to 1 m^2 or 10log (RCS/1 m^2)]. Note that these RCS diagrams will vary widely with the types of aircraft and ships involved. They will be significantly smaller for newer platforms designed for "stealth."

3.2.2 Radar Detection Range

To determine the range at which a radar can detect a target, it is necessary to consider one more value. That is the sensitivity of the radar receiver. The sensitivity is defined as the minimum signal level which the receiver can receive and still perform its specified functions (see Figure 3.9).

To determine the detection range, set the received power in any form of the radar range equation equal to the sensitivity and solve for the range. If we use the decibel form of the range equation, this is:

$$P_R = \text{Sens} = -103 + P_T + 2G - 20\log(F) - 40\log(d) + 10\log(\sigma)$$

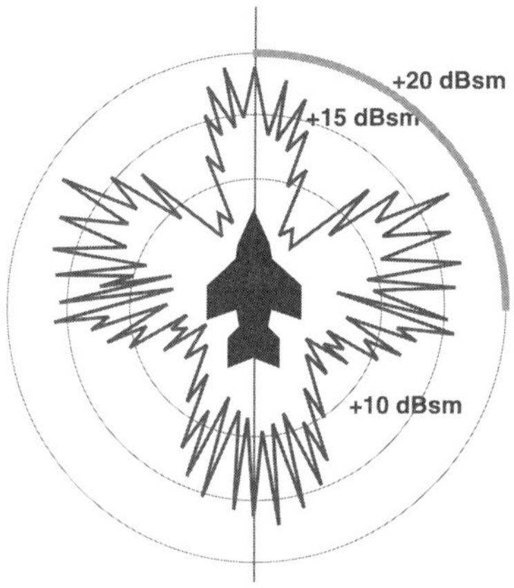

Figure 3.7 The radar cross section of an older aircraft is higher from the front and rear because the radar can "see" the engine parts. It is larger from the sides because of the larger cross section of the fuselage and the angles between the wings and the fuselage. Both of these effects are reduced in modern aircraft designed to

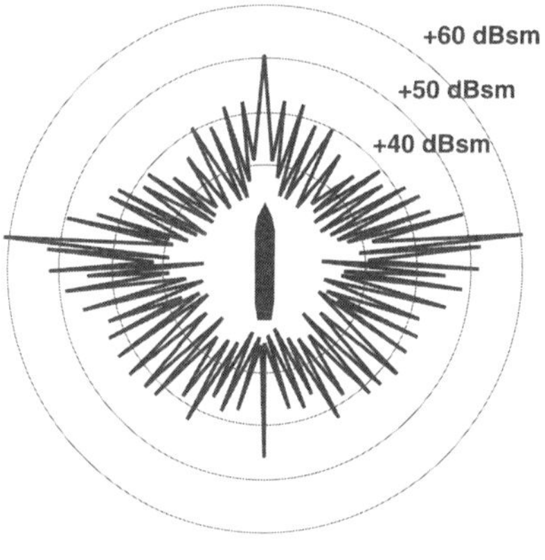

Figure 3.8 The radar cross section of a ship is typically starboard/port symmetrical. It may be characterized by very high RCS values 90° from the bow and lesser peaks fore and aft.

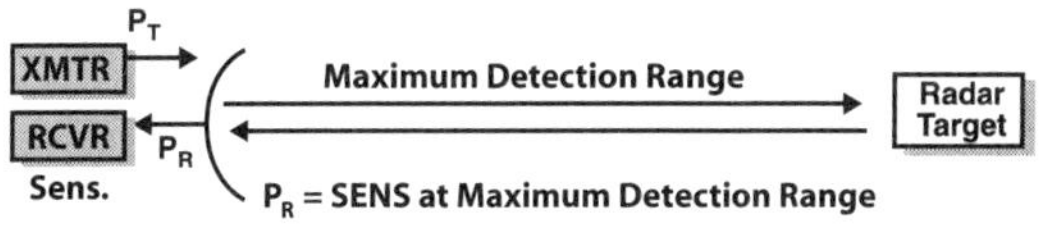

Figure 3.9 When the power received by the radar receiver is set equal to the sensitivity of the receiver, the radar range equation can be solved for the maximum detection range.

Then:

$$40\log(d) = -103 + P_T + 2G - 20\log(F) + 10\log(\sigma) - \text{Sens}$$

$$d = 10^{[40\log(d)/40]} = 10^{[(-103 + PT + 2G - 20\log(F) + 10\log(\sigma))/40]}$$

3.3 Detection Range Versus Detectability Range

A radar's detection range is the range at which it can detect a target. Its detectability range is the range at which its signal can be received and detected by an EW or reconnaissance receiver. Both of these range numbers are very situation dependent. The radar's detection range is a function of its parameters and the target's radar cross section.

To determine the radar's detectability range, we need to know: Is the receiver at the target or away from the target? What are the parameters of the receiver system detecting the radar?

As shown in Figure 3.10, the target is in the radar antenna main beam. The radar either tracks the target to keep it at the peak of its main beam or scans the beam through the target location. This means that the radar detection range equation, as described in Section 3.2, is applicable. Setting the received power equal to the sensitivity, we can solve for the range at which this occurs:

$$40\log(d) = -103 + P_T + 2G - 20\log(F) + 10\log(\text{RCS}) - \text{Sens}$$

where

d = the range from the radar to the target (in kilometers);

P_T = the radar's transmitter power (in dBm);

G = the radar's antenna gain (in decibels);

F = the transmitted frequency (in megahertz);

RCS = the radar cross section of the target (in square meters);

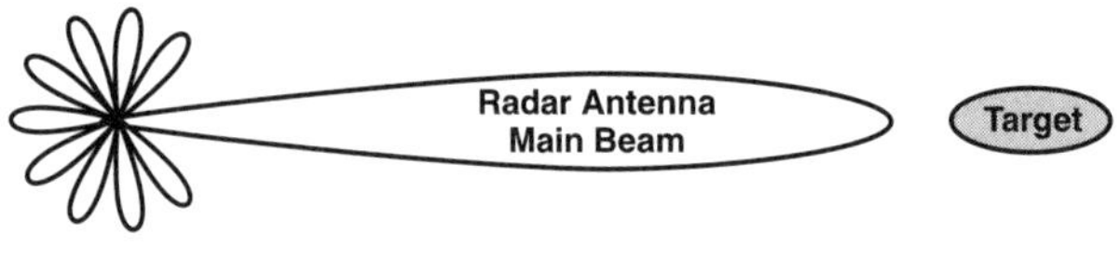

Figure 3.10 The peak gain point of the radar antenna either tracks the target or passes through its location during its scan.

Sens = the sensitivity of the radar's receiver (in dBm).

Then we find the range (d) from:

$$d = 10^{[40\log(d)/40]} \text{ or "antilog" of } [40 \log(d)/40]$$

With a scientific calculator, d is easily found by entering the value for 40 log(d), dividing by 40, then pressing "=," then second function, then log. However, unless we know (or estimate) the radar receiver's sensitivity we cannot determine the range.

3.3.1 Estimating the Sensitivity of the Radar Receiver

In an EW situation, the actual sensitivity of the radar's receiver is often not known—so it must be estimated in order to estimate the detection range of the radar. The sensitivity is defined as the lowest signal a receiver can receive and still do its job. High sensitivity means that the receiver can accept a very low signal level.

As shown in Figure 3.11, the sensitivity of any receiver is the product (that is, the sum in decibels) of kTB, the receiver noise figure, and the radar's required signal-to-noise ratio. kTB is the thermal noise in the receiver, which is determined from the receiver bandwidth by the equation:

$$\text{kTB} = -114 \text{ dBm} + 10 \log\left(\text{bandwidth}/1 \text{ MHz}\right)$$

If you know the radar's bandwidth, use that to calculate kTB. Otherwise, the bandwidth can be estimated from the radar's pulse width with the equation:

$$\text{Bandwidth} \cong 1/\text{pulse width}$$

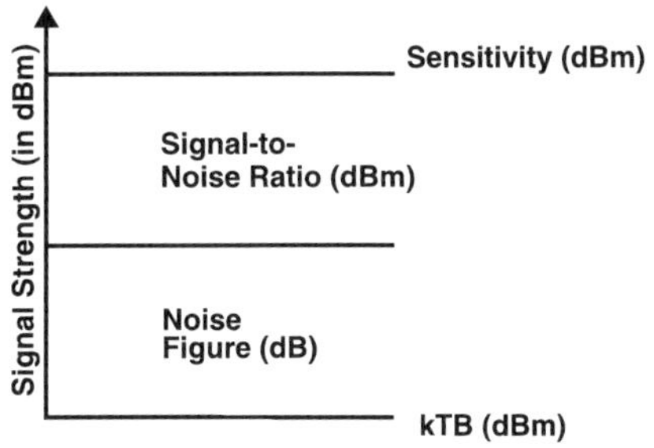

Figure 3.11 The sensitivity of a receiver is the sum (in decibels) of kTB, noise figure, and required signal-to-noise ratio.

The signal-to-noise ratio can be assumed to be ≈ 13 dB (a typical value) unless you know the actual required value. A typical value for the noise figure might be 5 dB.

For example, if the radar pulse width is 1 μs, the bandwidth would be assumed to be 1 MHz:

kTB is then –114 dBm

Taking the typical values for noise figure and signal-to-noise ratio makes the sensitivity:

$$\text{Sens} = -114 \text{ dBm} + 5 \text{ dB} + 13 \text{ dB} = -96 \text{ dBm}$$

3.3.2 Example of Radar Detection Range Calculation

This calculation uses the method shown above.

Let the sensitivity be –96 dBm, and let the other radar parameters be:

P_T = 100 kw (which is + 80 dBm);

G = 30 dB;

Frequency = 10 GHz;

Radar cross section of the target = 10 m^2.

Plugging these values into our expression for 40 log (d):

$$40\log(d) = -103 + 80\text{ dBm} + 2(30)\text{ dB} - 20\log(10{,}000)\text{ dB}$$
$$+10\log(10)\text{ dB} - (-96\text{ dBm}) = -103 + 80 + 60 - 80 + 10 + 96 = 63\text{ dB}$$

So d = antilog[40 log (d)/40] = antilog[1.575] = 37.6 km

3.3.3 Detectability Range

Now let's consider the range at which the radar signal can be detected by a receiver. We will take two cases. One is a radar-warning receiver (RWR) that is located on the target. The second is an ELINT (electronic intelligence) receiver that is located away from the target. Both of these cases are shown in Figure 3.12. We will determine the detection range for both cases and will compare each with the radar's detection range.

3.3.3.1 Detection Range for Radar-Warning Receiver

The RWR is designed to detect radars associated with threats as part of the process of protecting the target from those threats. It must detect a wide range of

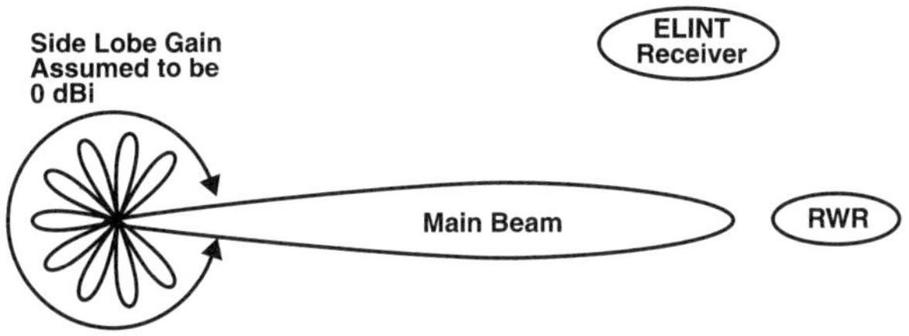

Figure 3.12 Since an RWR is located on the target, it can detect the radar antenna main beam. The ELINT receiver is typically required to detect radars in their antenna sidelobes.

radar signals and those signals can be coming from any direction. Since the peak of the main beam of the radar antenna is pointed at the target, the RWR will see the peak gain of the radar antenna. Because the RWR cannot be optimized to any specific radar, its bandwidth must be wide enough to accept the most narrow pulse widths expected. A typical RWR video bandwidth is thus 10 to 20 MHz. The RF bandwidth is typically 4 GHz, so if there is RF gain (some have it, some do not), the noise bandwidth will be a few hundred megahertz, determined by the formula:

$$BW_{EFF} = \mathrm{Sqrt}\left(2\, B_{RF}\, B_{VID}\right)$$

where

BW_{EFF} = the effective bandwidth;

B_{RF} = the RF bandwidth;

B_{VID} = the video bandwidth.

For example, with 4-GHz RF bandwidth and 10-MHz video bandwidth, the effective bandwidth is: BW_{EFF} = Sqrt(2 × 4,000 × 10) = 283 MHz.

Without RF gain, the video bandwidth is the effective receiver bandwidth. Since the signals can come from any direction, the RWR uses antennas with wide beamwidth. The antennas also have wide frequency coverage. The combination of these two factors dictate that RWR antennas have low gain (from about 2 dBi at the highest frequency to about –15 dBi at the lowest frequency). A typical RWR antenna will have approximately 0-dBi peak gain at 10 GHz. Since these antennas are used in combination, the effective antenna gain factor (at 10 GHz) for the RWR system will be 0 dBi for any direction of arrival.

The radar to RWR link is shown in Figure 3.13. The received power in the RWR is:

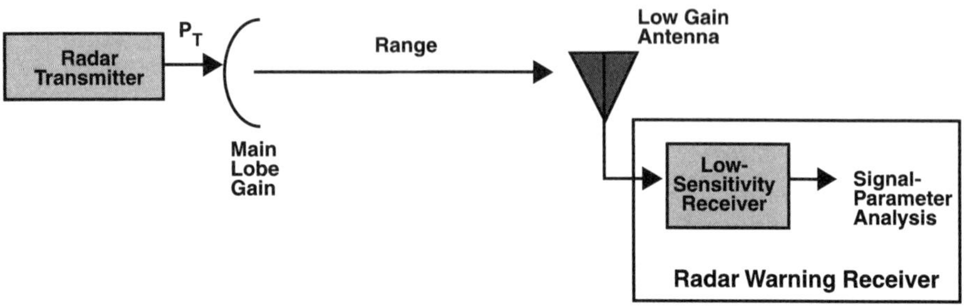

Figure 3.13 The RWR is typically designed to detect and identify a wide range of radar types in their main beams using low-gain antennas and low-sensitivity receivers.

$$P_R = P_T + G_M - 32 - 20\log(F) - 20\log(d) + G_R$$

where

P_R = the received power (in dBm);
P_T = the transmitted power (in dBm);
G_M = the main-beam peak gain of the radar antenna (in decibels);
F = the transmitted frequency (in megahertz);
d = the distance from the radar to the receiver (in kilometers);
G_R = the receiving antenna gain (in decibels).

To determine the detection range of the receiver, we set the received power equal to the receiver sensitivity and solve for the range.

$$P_R = \text{Sens} = P_T + G_M - 32 - 20\log(F) - 20\log(d) + G_R$$

$$20\log(d) = P_T + G_M - 32 - 20\log(F) + G_R - \text{Sens}$$

Then we determine d from:

$$d = 10^{[20\log(d)/20]} \text{ or "antilog" of } [20\log(d)/20]$$

The most common types of radar warning receivers use preamplified crystal video receivers which have approximately –65-dBm sensitivity. Using the previous values for the radar parameters, the detection range is:

$$20\log(d) = +80 + 30 - 32 - 20\log(10{,}000) + 0\text{ dB} - (-65) = 63\text{ dB}$$

$$d = \text{antilog}\left(63 / 20\right) = 1{,}413 \text{ km}$$

The ratio of the detectability range to the detection range is very large (≈37.6).

3.3.3.2 Detection Range for ELINT Receiver

An ELINT receiver will typically not be located in the main beam of the radar antenna. Thus the transmitter antenna gain is that of the radar antenna side lobes. It is common practice to assume that the side lobes of a narrow beam antenna are 0 dBi for older radars and up to 20 dB lower for many modern radar threats. A 0 dBi gain means the side lobes are lower gain than the main beam by an amount equal to the main-beam gain.

An ELINT receiver typically has a narrow-band receiver, so its sensitivity is calculated from kTB, noise figure, and required signal-to-noise ratio. Like the RWR, the ELINT receiver must accept a wide range of radar types, so its video bandwidth needs to be about 10 MHz. Note that most ELINT systems use superhet receivers which have wide front-end bandwidth; however, the bandwidth of each stage of a superhet receiver is typically narrower than the preceding stage. A general rule of thumb for the effective bandwidth of a superhet receiver is that it is approximately equal to the final prediction bandwidth (twice the video bandwidth for AM detection). Thus, kTB is –104 dBm [–114 + 10 log(20)] = –101 dBm. The noise figure and required signal-to-noise ratio should be approximately the same as those of the radar (which we set at 10 dB and 13 dB, respectively), so the sensitivity of a typical ELINT receiver would be:

$$\text{kTB} + \text{NF} + \text{SNR} = -101 \text{ dBm} + 10 \text{ dB} + 13 \text{ dB} = -78 \text{ dBm}$$

The ELINT receiver system could reasonably have an antenna with moderate gain—perhaps 10 dB. This makes the effective range equation (as derived earlier for the RWR case):

$$20\log\left(d\right) = P_T + G_S - 32 - 20\log\left(F\right) + G_R - \text{Sens}$$

where G_S is the radar antenna side lobe gain (which we assumed to be –10 dB)

$$20\log\left(d\right) = +80 - 10 - 32 - 20\log\left(10{,}000\right) + 10 - \left(-78\right) = 46$$

$$d = \text{antilog}\left(46 / 20\right) = 200 \text{ km}$$

Since the radar detection range (calculated earlier) is 37.6 km, the ratio of the detectability to detection ranges for this case is ≈ 5.3.

3.4 Radar Modulation

The modulation on a radar signal allows the radar to measure the range to a target which reflects the radar's transmitted signals. Since electromagnetic signals propagate at approximately the speed of light (about 3×10^8 m/s) the distance from the radar to the target is determined by measuring the round-trip propagation time. The distance is half of the time that the received signal is delayed from the transmitted signal multiplied by the speed of light (see Figure 3.14).

A very practical problem is to determine when the signal leaves and when the return is received. Since a single-frequency CW signal repeats once every wavelength (much less than a meter for microwave signals), it is useless for measuring propagation delays associated with radar/target engagements. However, modulation on the signal, which has significantly lower frequency, provides something to measure and compare over the necessary time intervals (of the order of milliseconds).

There are many types of modulations used, but they fall into the categories of pulse, linear FM, binary modulation, and noise (or pseudonoise) modulation. Unmodulated CW can measure relative velocity between the radar and the target, which can be extremely useful, but we'll consider that a little later.

3.5 Pulse Modulation

A pulse is just a very short signal transmission with reasonably clean turn-on and turnoff characteristics as shown in Figure 3.15. In its basic form, the pulse has fixed radio frequency and is characterized by its pulse width (or pulse duration) and pulse repetition interval (or pulse repetition frequency). The duty cycle (pulse duration/pulse interval) is relatively low. The pulse provides a clearly measurable time event in the signal. The measured event may be the whole

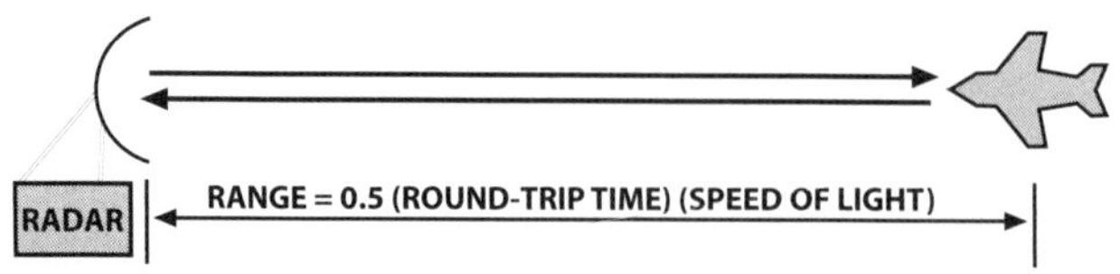

Figure 3.14 The radar measures the distance to a target as a function of the round-trip propagation time.

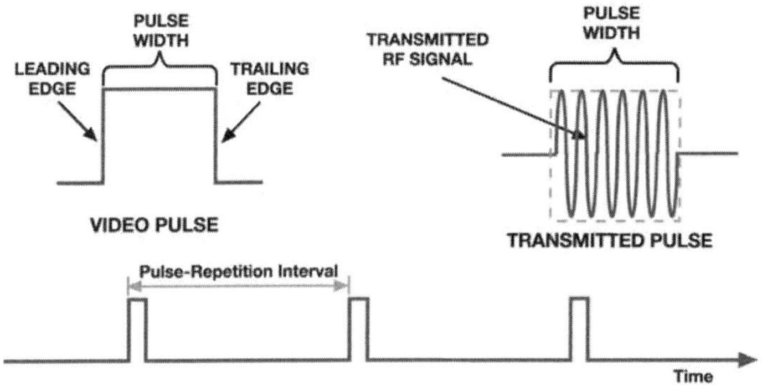

Figure 3.15 The pulse (video) turns an RF transmitter on for the duration of the pulse and repeats it at the pulse-repetition interval.

pulse, or (if the radar's receiver has sufficient bandwidth) it may be the leading edge of the pulse. In either case, the time from the transmission of a pulse to the receipt of a reflected pulse is easily measured.

Pulse radars have the significant advantage that their receivers are turned off during pulse transmission. This allows the radar to use a single antenna for transmission and reception and protects the receiver from both saturation and damage.

The pulse repetition rate determines the maximum range at which the radar can make unambiguous range measurements as shown in Figure 3.16. If a second pulse is transmitted before the reflection of the first pulse from a target reaches the radar, the delay time measurement would start with the second transmitted pulse and end with receipt of the first pulse. Thus the

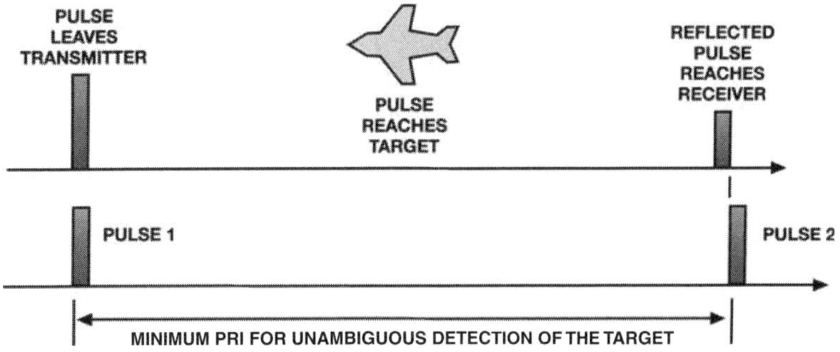

Figure 3.16 The PRI of a radar signal limits the unambiguous range, because each pulse reflection must be received before another pulse is transmitted.

round-trip propagation time is not accurately measured (assuming that the pulses are identical).

The pulse duration determines the minimum range at which the radar can detect a signal. The receiver is not turned on until the trailing edge of the pulse leaves the transmitter (plus some guard time). The reflected leading edge of the pulse cannot reach the receiver before the trailing edge has been transmitted.

The pulse width also determines the range resolution of the radar—that is the range difference between two targets which will allow the radar to determine that there are two targets. This is shown in Figure 3.17. Consider the pulse in the vicinity of two targets. The round-trip distance between the first and second targets must be greater than the pulse width for the receiver (and processor) to be able to separate the two returns.

3.5.1 Unintentional Modulation on Pulses

A closer look at a radar pulse will show that it has a rise time and a fall time. The rise time is the time required for the pulse to go from 10% of its transmitted power to 90% of its power. The fall time is the opposite (for the trailing edge). There may also be ringing, or other unintentional modulation effects including unintentional frequency modulation). While these effects are important to EW

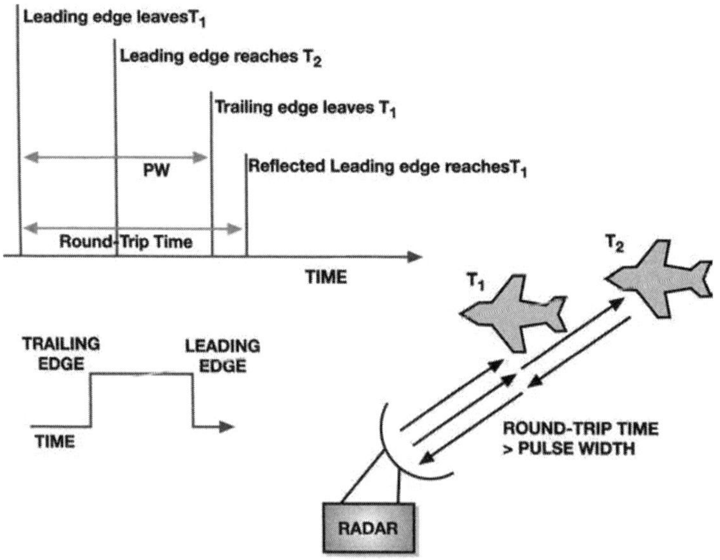

Figure 3.17 The round-trip propagation time between two targets must be greater than the pulse duration to allow the radar to detect the presence of two distinct targets.

systems which perform specific emitter identification (SEI), they do not impact the basic function of the pulse to the radar.

3.5.2 Pulse Compression

Pulse compression is a way to increase the range resolution of long-pulse radars. The effect of compression is shown in Figure 3.18. Note that compressed radars are used for long-range detection, so they require high-energy pulses. The peak power is made as high as practical, then the pulse energy is increased by the large pulse width. The detectability of the radar is a function of the transmitted peak power, but its detection range is a function *of the total transmitted energy that is returned from the target.* The long pulse is reflected by the target, but the range resolution is improved by the shortened pulse from the compression function in the receiver. There are two important techniques for achieving pulse compression. One involves the addition and processing of frequency modulation and the second involves the addition and processing of a digital modulation.

3.5.3 Chirped Pulses

A pulse with a linear frequency modulation, as in Figure 3.19, is said to be "chirped." The frequency of the pulse can either increase or decrease with time, but the figure shows an increasing frequency modulation. The frequency-modulated pulse is transmitted and received just like a fixed-frequency pulse, but in the receiver, it passes through a compressive filter. The compressive filter causes a delay that is a function of frequency—the higher the frequency, the less the delay. The delay versus frequency function is linear, and matched to the modulation placed on the pulse. The difference between the maximum and

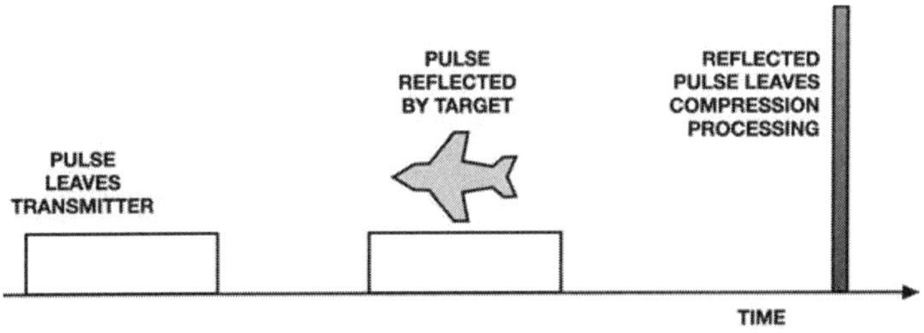

Figure 3.18 The pulse is transmitted with a reasonable peak power and is reflected from the target. However, by compressing the received reflection from the target, the radar performance is as though the transmitted power were greater and the pulse duration shorter.

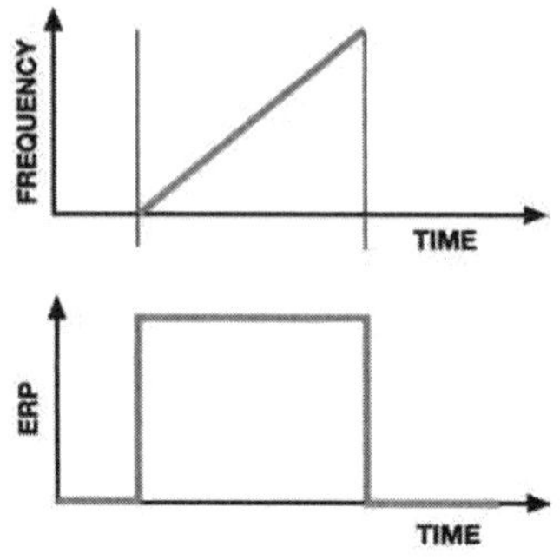

Figure 3.19 A chirped pulse has a linear frequency modulation during the pulse duration.

minimum delay is equal to the pulse width. The function of the compressive filter is shown in Figure 3.20.

Note that the pulse from the compressive filter has all of the received energy concentrated in a time period much smaller than the transmitted pulse width. The transmitted pulse width is labeled "A" in the figure and the effective pulse width after compression is labeled "B." The ratio of A to B is the compression factor. Chirped radars sometimes have very large compression factors. Since the depth of the radar's resolution cell (i.e., its range resolution) is half of the received pulse width, the range resolution is improved by the compression factor. The radar resolution cell is the area within which the radar cannot distinguish multiple targets.

The detection range for any given target will remain the same because the power reflected by the target remains the same. This may be confusing to EW people who are used to having the intercept range proportional to the square

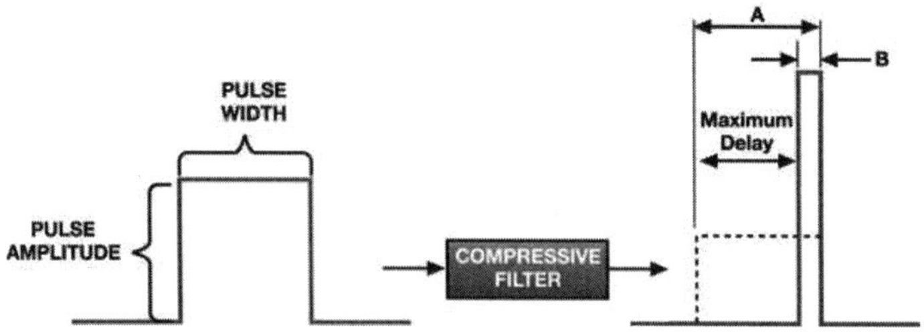

Figure 3.20 The frequency versus delay slope of the compressive filter delays each portion of the pulse to the end of the pulse, which created a much shorter pulse with the same total energy as the prefiltered pulse.

root of the transmitted peak power. A way to think about this is that the compressed pulse is narrower, requiring more bandwidth. The increased bandwidth raises the sensitivity threshold by the amount that the pulse peak power is increased by the compression. Ignoring losses in the compression process, the detection range for any given target will increase by the fourth root of the compression ratio. (Because the received power is a function of the fourth power of range (range^4 or 40 log[range].)

3.5.4 Digital Modulation on Pulses

Another way to increase the radar range resolution relative to that derived from the pulse width is to apply a digital modulation to the pulse. In Figure 3.21, the pulse has a seven-bit pseudorandom code applied as a BPSK modulation. The phase of the RF signal is shifted 180° during the bits denoted as "–" while the bits denoted as "+" remain at the reference phase. After decoding, the effective pulse width will be the bit duration rather than the pulse duration.

The tapped delay line assembly is shown in Figure 3.22. The delay line is tapped with the taps separated in time by the bit period. There are as many taps as the bits in the modulating code, and the delay line is as long as the pulse. The signals from all of the taps are summed to form the output. Note that if the pulse exactly fills the delay line, the bits in the pulse that are phase shifted are passed through 180° phase shifters before summing with the other tap outputs. The bottom part of Figure 3.22 shows 13 snapshots of the pulse passing through the delay line. On the first line, only the first bit has entered the delay line, and on the thirteenth, only the last bit is still in the delay line. To the right of each bit sequence, the plus and minus values coming from the taps are added to form the output value. Note that only the bits that are in the delay line are summed at the output. Each position has a total value of either 0 or –1 except the position

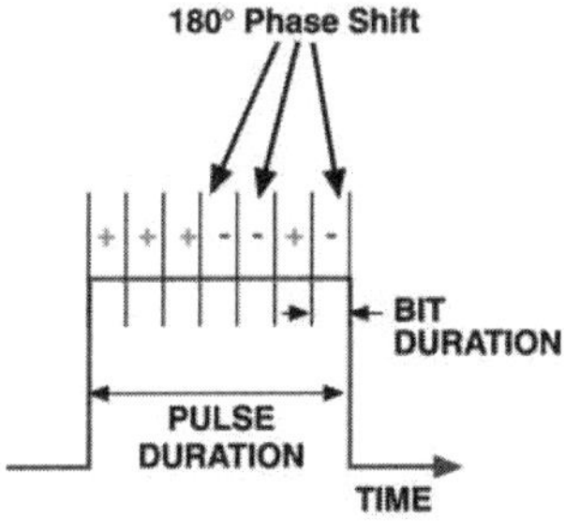

Figure 3.21 A binary phase shift modulation can be added to the pulse to provide compression. "+" indicates no phase shift and "–" indicates 180° phase shift in this binary phase shift keyed modulation.

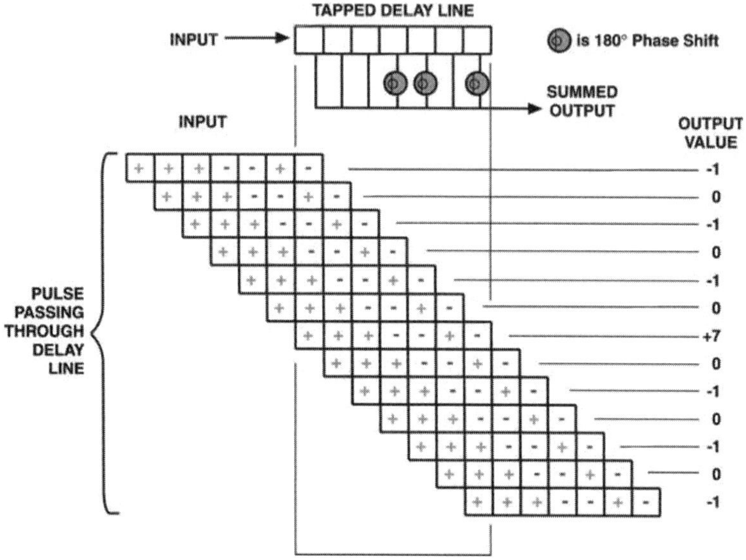

Figure 3.22 The receiver has a delay line with taps at intervals equal to the bit duration and a total length equal to the pulse duration. As the pulse passes through the delay line, the sum of the taps is a very low number, unless the pulse exactly fills the delay line, in which case the output has a strong peak.

in which the pulse exactly fills the shift register. In this position, the value is +7. The range resolution is improved by a factor equal to the number of bits in the code.

Figure 3.23 shows the summed output of the delay line as a function of time as the pulse moves linearly through it. This demonstrates the so-called "thumb tack" correlation factor for pseudorandom binary coded signals. You will be seeing this again when we talk about low probability of intercept radar modulations later. Note that the correlation starts increasing linearly when the code comes within one bit of alignment. It reaches a sharp peak when the bits are completely in phase with each other. Then, it ramps down linearly until the signal is again 1 bit out of synchronization.

3.6 CW and Pulse Doppler Radars

If a radar uses coherent signals, the Doppler principle allows it to determine the rate of change of distance to a target. This allows the radar to separate the signal reflected from a moving target from ground reflections. Being able to

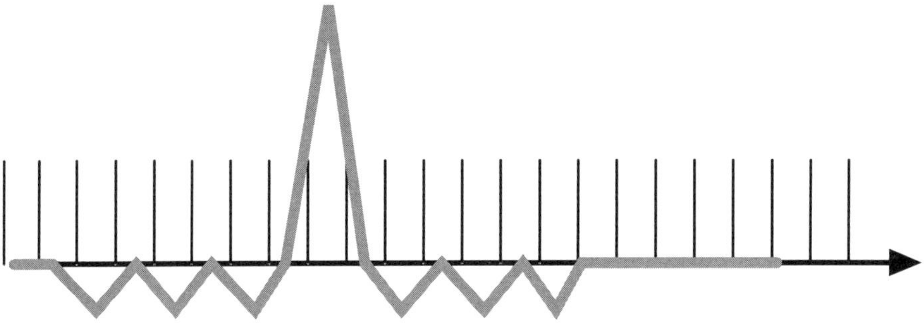

Figure 3.23 This is the correlation function of the pulse passing through the delay line. As with any digital signal, it has a "thumbtack" correlation function.

distinguish a target against the ground gives a radar-controlled weapon system a look down/shoot down capability.

3.6.1 Doppler Shift

A signal propagating from a moving transmitter to a fixed receiver will be received with a difference frequency from the transmitted signal determined by the formula:

$$\Delta F = \left(v/\mathrm{c}\right)F$$

where:

ΔF = the change in received frequency from the transmitted frequency;

v = the component of the transmitter velocity in the direction of the receiver;

c = the speed of light;

F = the transmitted frequency.

Since the radar return signal has made a round trip, it has twice the frequency shift. Also, since both the radar platform and the target may be moving, a general expression for the Doppler shift in a radar return is:

$$\Delta F = 2\left(V/\mathrm{c}\right)F$$

where V is the instantaneous rate of change in the distance between the radar and the target, and all other definitions are the same.

An interesting side note here is that a defensive tactic used in air combat is "notching" by turning the defending aircraft so that its flight path is perpendicular to the direction of a weapon radar—causing the Doppler shift to drop to zero.

3.6.2 CW Radar

True CW radar cannot measure the range to a target except by measurement of the return signal power which is not very accurate. However, it does determine the range rate by measuring the Doppler shift. As shown in Figure 3.24, a CW radar must usually have separate transmitting and receiving antennas to avoid excessive leakage of power from the transmitter into the receiver. This is because both the transmitter and receiver are on at the same time. The receiver must use a common frequency reference with the transmitter in order to measure the Doppler shifts—which are very small compared with the transmitted frequency. For example, a 10-GHz radar would see about 18.5 Hz of Doppler shift for each kilometer/hour of closing speed. Rules of thumb given by George Stimson (see reference in Appendix C) are shown in Table 3.1.

3.6.3 FM Ranging

In order to accurately measure distance to the target, a linear frequency modulation ramp can be applied to the transmitted signal as shown in Figure 3.25. This modulating signal can either have a fixed-frequency portion or it can be a two-directional frequency ramp.

First, consider the linear ramp portion of the modulating waveform in Figure 3.26. The received signal is delayed from the transmitted signal by the

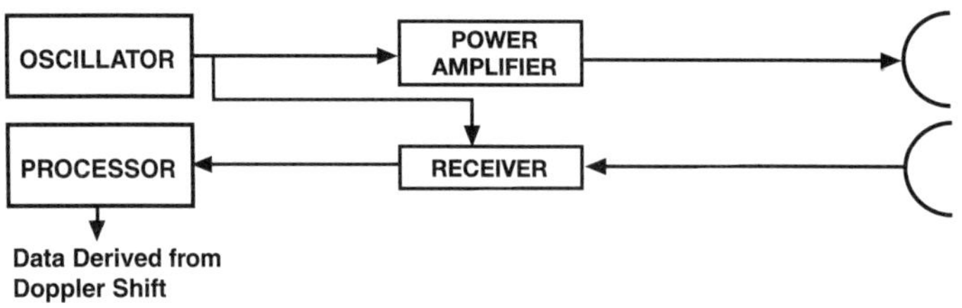

Figure 3.24 A CW radar generally requires separate transmitting and receiving antennas to prevent leakage of transmitter power into the receiver. It can only determine the range rate to the target by comparing the transmitted and received signal frequencies.

Table 3.1
Doppler Frequency for X-Band Radars

Rate of Change of Distance	Doppler Frequency
1 nautical mile/hour	35 Hz
1 mile/hour	30 Hz
1 kilometer/hour	19 Hz
1,000 feet/second	20 kHz

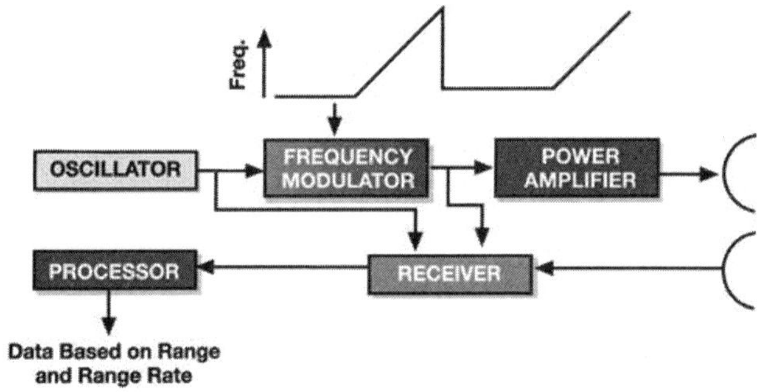

Figure 3.25 If the CW radar signal has a linear frequency modulation, the transmitted and received signals can be compared to provide both range rate and actual range to the target.

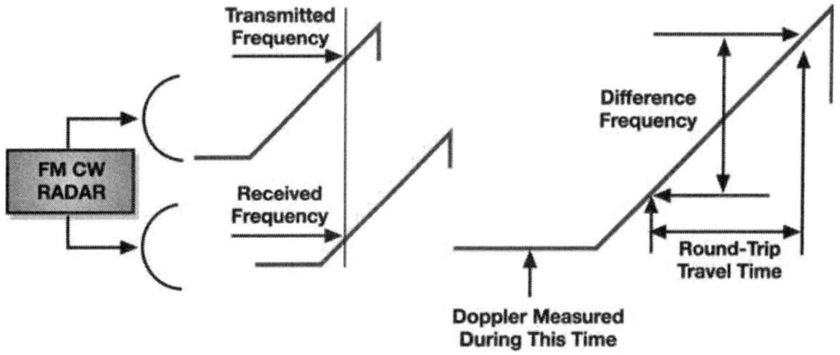

Figure 3.26 The round-trip propagation time between the radar and the target causes a difference between the transmitted and received signal frequencies that is dependent on the frequency modulation versus the time slope.

amount of time it took the signal to reach the target and return (at the speed of light). Thus, by comparing the transmitted and received signals at any instant during which both the transmitted and received signals are in the linear ramp portion of the modulating waveform, the distance can be measured. This is shown at the right side of the figure.

Now, consider that the difference frequency measured is actually caused by two factors—the round-trip propagation time and any (positive or negative) Doppler shift caused by the rate of change of distance. If the modulating waveform has a constant frequency portion, the Doppler shift can be measured during that part of the signal and the range measurement adjusted accordingly.

If the radar uses a bidirectional waveform, the range-related frequency shifts will have opposite senses during the up and down frequency ramps—while the Doppler shift will be in the same direction. This will allow the Doppler component to be measured and the range accurately calculated.

3.6.4 Pulse Doppler Radar

As shown in Figure 3.27, the pulse Doppler radar in its high PRF mode outputs a coherent pulsed signal with a high duty factor. This pulse train also has an extremely high PRF which makes it challenging to radar-warning receiver processing. Coherent pulses are formed by interrupting a continuously running oscillator, so each received RF pulse will be in phase with an oscillator that is phase locked to the RF waveform of all previous pulses. This allows the advantages of synchronous detection and also allows Doppler shifts to be measured.

Since the receiver is turned off during pulse transmissions, a single antenna can be used without the very difficult isolation problems associated with CW radars.

The radar can measure range just like any other pulse radar, but it will have significant blind ranges and range ambiguities. These can be resolved by use of multiple FM ranging or the use of other operating modes, and the application of multiple pulse repetition rates implemented in sophisticated processing.

3.7 Moving Target Indicator Radars

The moving target indicator (MTI) is a type of radar designed to detect moving targets on the ground. It does this by sensing the Doppler shift of detected targets. MTI radars can either be ground-based or airborne. The airborne MTI (sometimes called AMTI) has added complexity because the radar itself is moving, causing additional Doppler shifts.

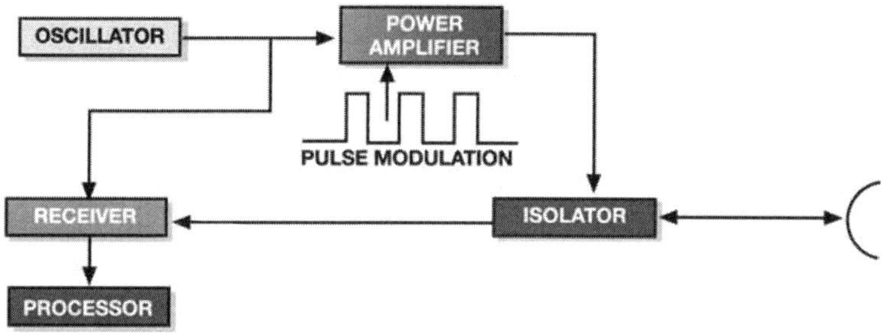

Figure 3.27 A pulse Doppler radar outputs a coherent pulse stream at a very high duty factor. The receiver is off during transmissions, relieving the leakage problem in a single antenna. Range to the target can be determined either by pulse timing or frequency modulation. Range rate is determined from the Doppler shift of return signals.

3.7.1 Basic MTI Operation

The MTI sweeps an antenna over an angular segment (up to 360°) and covers a specified range. It determines the presence of moving targets in cells as shown in Figure 3.28. The angular resolution is derived from the scanning of the antenna beam, and the range resolution is from the return of pulses from everything that reflects the pulses.

Like any radar, the range resolution is set by the pulse width—which is typically very narrow. The pulses may be chirped to improve the range resolution (i.e., reduce the depth of the range resolution cells). If pulse compression is used, the compressed pulse is processed just like the noncompressed pulse.

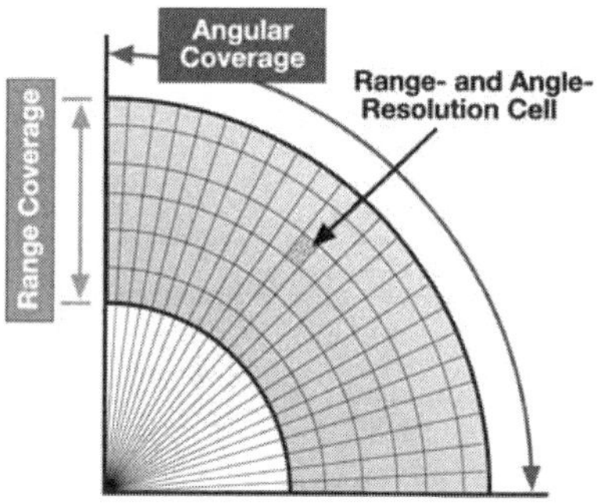

Figure 3.28 An MTI radar determines the presence of moving targets within range- and angle-resolution cells.

Since the transmitted pulse and any reflected return propagate at the speed of light, the reflected pulse arrives at the radar with a delay from the transmitted pulse of:

$$\left[2d/\left(3\times10^{8}\right)\right]\text{ seconds}$$

where *d* is the range to the reflecting object in meters.

As shown in Figure 3.29, the MTI samples the return signals once during each pulse width. This sampling can continue during the whole time interval until another pulse is transmitted. Sampling thus looks for return energy from range increments which equal 1m per 6.6 ns of pulse width (or of compressed pulse duration).

At each sample, an A/D converter makes "in-phase and quadrature" (I and Q) digitizations of the signal. Since these two digital words represent points on the received waveform that are 90° apart in phase, they allow the determination

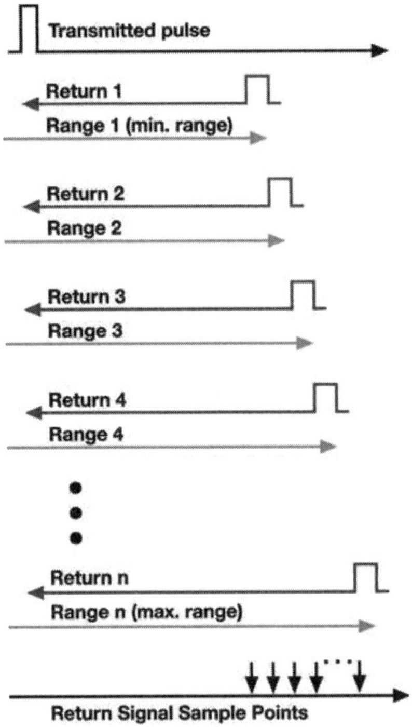

Figure 3.29 An MTI radar transmits a narrow pulse and samples the return from that pulse at intervals equal to the pulse width over a period that accepts returns from the minimum to the maximum range.

of the received frequency and phase. This process is continued (for each pulse) over the whole range of interest.

This sampling pattern is repeated for each pulse. Then the values for each sample from pulse 1 are subtracted from the equivalent sample values of pulse 2 as shown in Figure 3.30. The values from pulse 2 are subtracted from those of pulse 3, and so on through pulse $m-1$ and pulse m. The m pulses illuminate each angular resolution cell during each sweep of the antenna. More complex data subtraction schemes are sometimes implemented to provide better clutter cancellation.

Now, all of the "sample 1" measurements for the m pulses are used to generate a fast Fourier transform (FFT). The FFT determines the presence of Doppler shifted signals in each range and angle resolution cell.

The Doppler shift is caused by the rate of change of distance between the radar and the target. Thus, the MTI can only detect targets which have some component of motion directly toward or away from the radar.

3.7.2 MTI Data Rates

MTI radars generate a great deal of raw data. For example, an MTI radar with pulse repetition frequency of 6,250 which samples 200-times-per-pulse repetition interval and digitizes the I and Q values at 12 bits each—generates 30 Mbps of raw data.

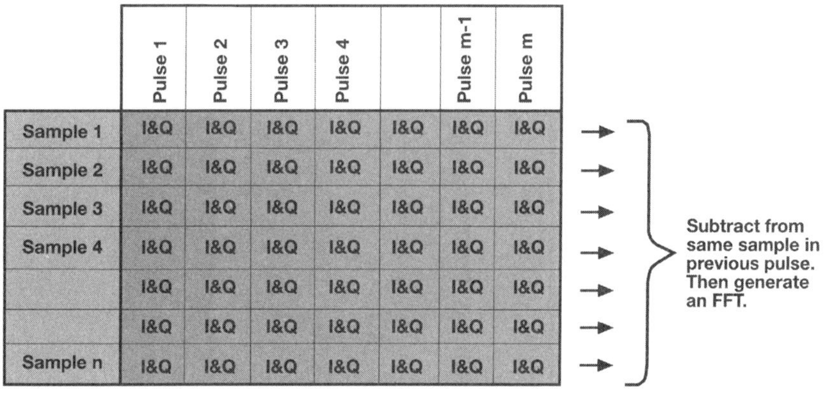

Figure 3.30 I&Q samples are collected for each sample point for each pulse that illuminates a moving-target resolution cell. Then each sample is subtracted from the equivalent sample in the previous pulse, and an FFT is calculated from the resulting differences for all pulses illuminating from the cell.

However, the MTI processing reduces this data to a much more manageable level for display or reporting. Since the MTI only reports the presence and magnitude of motion in resolution cells, each target report need only contain the cell location and the magnitude and sense of the movement; 80 bits is typically plenty of data for each target. If there are 100 moving targets detected per second in the covered area, the total target report data rate will be 8 Kbps. Even adding a 64-bit status word 30 times per second yields a total output data rate of less than 10 Kbps. This data rate is easily carried over an audio bandwidth link.

3.7.3 Airborne Moving Target Indicator (AMTI) Radar

If the MTI radar is mounted in an aircraft, it operates just like the basic MTI described earlier, except that it has the additional problem of resolving Doppler shifts caused by the aircraft's motion. As shown in Figure 3.31, this Doppler shift is a function of both the aircraft's speed over the ground and the angle of the radar antenna relative to the aircraft's ground track. The aircraft in the figure has been deliberately drawn to show that it is not the airspeed of the aircraft, but the ground speed that determines the induced Doppler shift. When the antenna is less than 90° from the direction of travel (over the ground), the aircraft induced Doppler shift is positive; beyond 90°, it is negative.

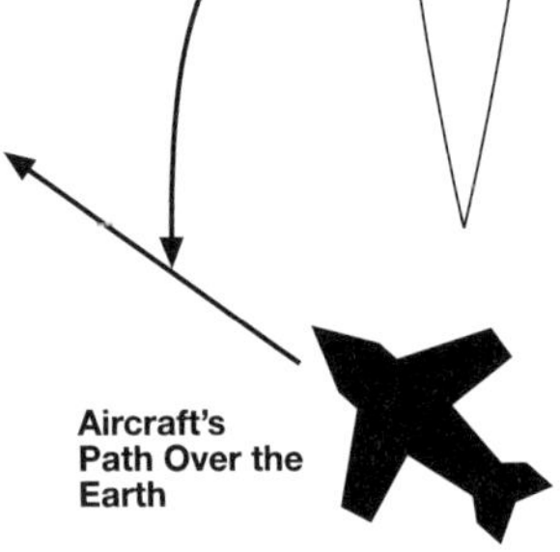

Figure 3.31 In an AMTI radar, the Doppler shift caused by the movement of the aircraft over the ground is subtracted from the Doppler shift in each cell before declaring and reporting moving targets.

As the antenna sweeps, it makes a changing angle (θ) with the direction of the aircraft's ground track. The aircraft-induced Doppler shift will be:

$$2FS\cos(\theta)/c$$

where

F is the radar frequency;

S is the aircraft ground speed;

θ is the angle between the ground track and the antenna boresight;

c is the speed of light (3×10^8 m/s).

The Doppler shift observed by the MTI radar in each resolution cell must be corrected for this aircraft Doppler before the presence of a moving target is reported. This can be accomplished by shifting the zero frequency point of the Doppler up or down by the above-stated amount. It can also be implemented by varying the local oscillator in the receiver or varying the transmitted frequency by the same amount.

3.8 Synthetic Aperture Radars

A synthetic aperture radar (SAR) takes advantage of the movement of an airborne platform to create, in effect, a phased array that is extremely long. This allows extremely high resolution at long range, with relatively small physical antennas. SARs are used to create maps of large areas, along with vehicles and other objects present in the area. There are combined MTI and SAR radars in which moving objects are identified and, as soon as the object stops moving, a SAR image of that object is made to allow it to be identified.

The SAR creates range and azimuth resolution cells (as shown in Figure 3.32) to the required resolution. These cells are often pictured as rectangles, but are in fact "blobs" of the indicated size since they are at the limit of resolution of the SAR. The required resolution is a function of the smallest objects that must be located or identified.

3.8.1 Range Resolution

Range resolution is determined by the pulse width of the radar. If it uses range compression (chirp or phase coding) the compressed pulse width determines the range resolution.

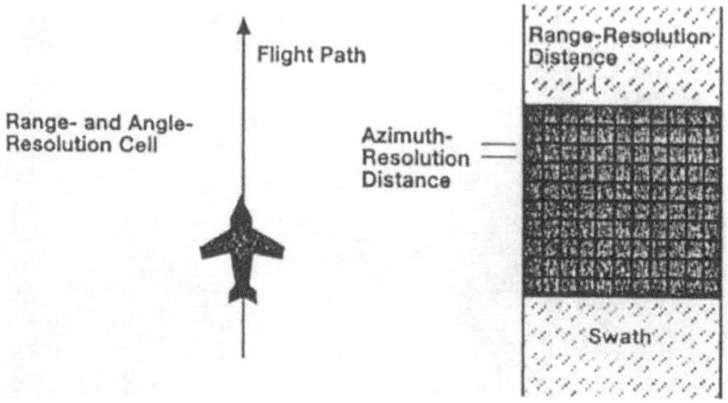

Figure 3.32 A typical SAR creates range- and azimuth-resolution cells in a swath parallel to the flight path of the aircraft carrying the radar.

$$d_r = c(PW/2)$$

where

d_r = the range resolution in meters;

c = the speed of light (3×10^8 m/s);

PW = the radar pulse width (or its compressed pulse width).

Like the MTI radar, the SAR measures returned signal power at time increments equal to the pulse width to create a series of "range bins" which determine the radar return as a function of distance from the radar as shown in Figure 3.33. I and Q samples are collected for each cell since the SAR process requires that the phase be preserved.

3.8.2 Azimuth Resolution

The azimuth resolution of a radar depends on the beamwidth of its antenna. The beamwidth is a function of the size of the antenna. For a parabolic dish antenna, the surface of the dish (actually a parabolic section) reflects all of the energy it receives to the feed (which is at the focus of the parabola). The larger the dish, the more narrow the antenna beam. For a phased array antenna, delay lines are used to create a coherent addition of signals received by many array elements (antennas) when those signals arrive from a single direction—thereby forming a narrow antenna beam. The longer the array, the more narrow the beam.

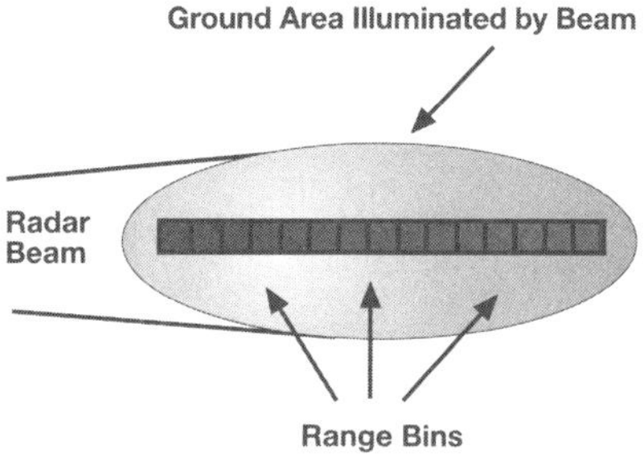

Figure 3.33 Range bins are formed along the boresight of the antenna beam by sampling the radar return at intervals equal to the pulse width.

Now, let's consider a simple SAR with an antenna mounted perpendicular to the flight path. The SAR transmits coherent pulses and creates the effect of a phased array by collecting the returns from each pulse as the platform moves forward. The distance that the aircraft moves between pulses is in effect the distance between the "antennas in the array" (i.e., with 300-meters-per-second aircraft speed and 300-pulses-per-second PRF, a pulse is transmitted every meter). Assuming that the area being mapped by the SAR is much farther from the aircraft than the flight distance over which data is collected, the returns from objects on the antenna boresight can be added in phase—while objects away from the boresight will add up out of phase. Thus, summing the data in corresponding range resolution cells over several pulses has the same beam-narrowing effect as a phased array.

Figure 3.34 summarizes the way SAR data is processed to form the synthetic array length. After each integration period, data from a new pulse is added and the data from the oldest pulse is discarded. It should be noted that the azimuth resolution (i.e., the cross-range resolution) achieved by a synthetic array of a particular length is half of the azimuth resolution achieved by a phased array of real antennas. The synthetic array azimuth resolution equation is:

$$d_a = \lambda R / 2L$$

where

d_a = the azimuth resolution distance (same units as R);

λ = the wavelength of the radar signal;

L = the length of the array (same units as (λ);

R = the range to the target.

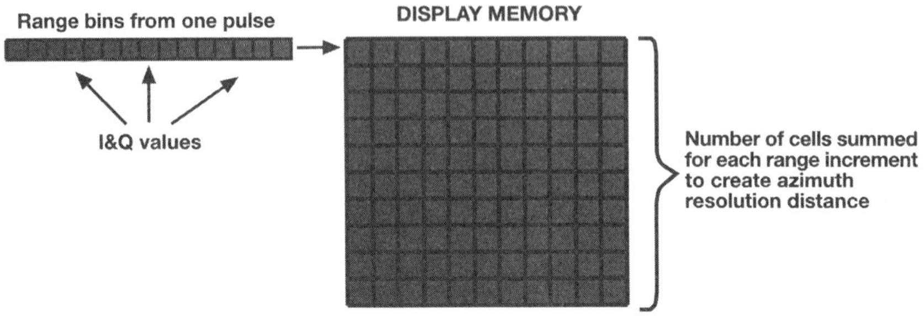

Figure 3.34 I&Q samples are collected for each range bin from each pulse. The data from the same range bin is summed for several pulses to form the azimuth-resolution distance.

For a phased array of real antennas, the azimuth resolution distance is:

$$d_a = \lambda R / L$$

3.8.3 Focused Array SAR

The earlier discussion assumed that an "unfocused array" was used. This requires that the paths from the target to the radar be very close to parallel for all of the integrated pulses. This limits the length of the synthetic array and thus the azimuth resolution. Much longer synthetic arrays can be formed using focused array techniques.

As shown in Figure 3.35, the phase difference between signals received from different pulses can be significant if the synthetic array length is significant relative to the range. The phase error is:

$$\phi_n = 2\pi d_n^2 / \lambda R$$

where

ϕ_n = the phase error for thc measurement made a distance of d_n from the point of closest approach to the target;

λ = the wavelength of the radar signal;

R = the range at the closest approach to the target.

In a focused array, this phase error is corrected before the azimuth data is summed for each range cell. This can require an immense amount of processing, but the processing load can be reduced by use of Doppler filters formed by an FFT.

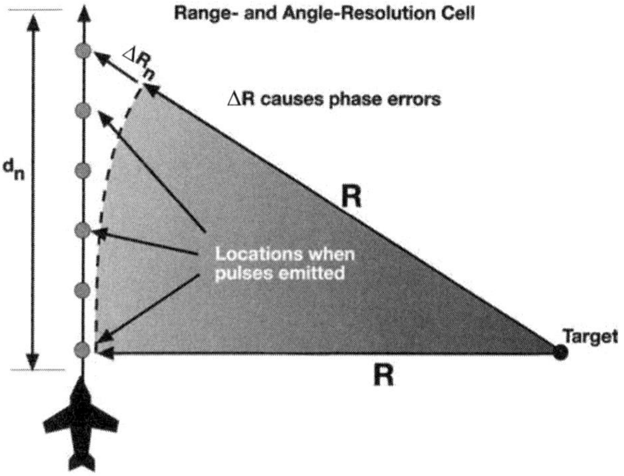

Figure 3.35 If the synthetic array is long, there can be significant phase errors in the target returns of some pulses.

3.9 Low Probability of Intercept Radars

The purpose of low probability of intercept (LPI) radar is to detect and track targets without being detected by electronic warfare receivers. Thus, an LPI radar is one that satisfies this very broad criteria. Whether or not a radar is LPI depends on what the radar is trying to do, what kind of receiver is trying to detect it, and the applicable engagement geometry. For purposes of this discussion, we will call the intercepting receiver system an ESM receiver. Table 3.2 contains a few definitions related to LPI radar.

3.9.1 LPI Techniques

A number of measures can be taken to make a radar less subject to detection. One is to make the signal so weak that the ESM signal cannot receive it. This is difficult for the radar because the radar must receive enough energy after the round trip to the target (40 log range in the radar range equation) to detect the target. The receiver encounters only a one-way path loss (20 log range). A second way is to narrow the radar beam (thus increasing the antenna gain) or to suppress antenna side lobes. This makes it more difficult for a receiver not located at the target to intercept the signal, but does not impact a receiver located on the target.

A third way to reduce the interceptability of a radar relative to its performance is to give the radar a processing gain not available to the ESM receiver.

Table 3.2
LPI-Related Definitions

Term	Definition
Coherent radar	Transmitted signals have a constant phase relationship to an oscillator in the transmitter.
Frequency agile radar	Each pulse or groups of pulses are transmitted at different frequencies.
LPID radar	A radar with parameters that make it difficult for an ESM receiver to correctly identify the radar type.
Quiet radar	A radar that detects a target at the same range that the target can detect the radar's signal.
Random signal radar	A radar which uses a waveform that is truly random (for example, noise).
Binary phase coded CW radar	A radar which has a pseudorandom phase coded modulation on a transmitted CW signal.

3.9.2 Levels of LPI

Radars can be thought of as having three levels of LPI:

- The radar is easily detectable but not easily identifiable—called an LPID radar (see Figure 3.36).
- The radar can detect a target and is not detectable by an ESM receiver at the same range but outside its main beam (see Figure 3.37).
- The radar can detect a target and is not detectable by an ESM receiver located on the target—a "quiet radar" (see Figure 3.37).

The ability of a receiver to detect the radar signal depends on its noise figure and its bandwidth. In the following analysis, we generally assume that the noise figure of the radar receiver and the intercepting receiver are the same, and that the intercept receiver bandwidth can be optimized to its function. Here we use ESM receiver as a general term to cover aircraft radar-warning receivers, shipboard ESM receivers, and ground-based warning and targeting receivers.

3.9.3 LPID Radars

As shown in Figure 3.36, an ESM receiver identifies a threat radar based on its parameters. The typical ESM processor has a threat identification (TID) table with a set of parameters for each expected threat signal type in each of its operating modes. The processor also tries to discriminate against friendly radars and

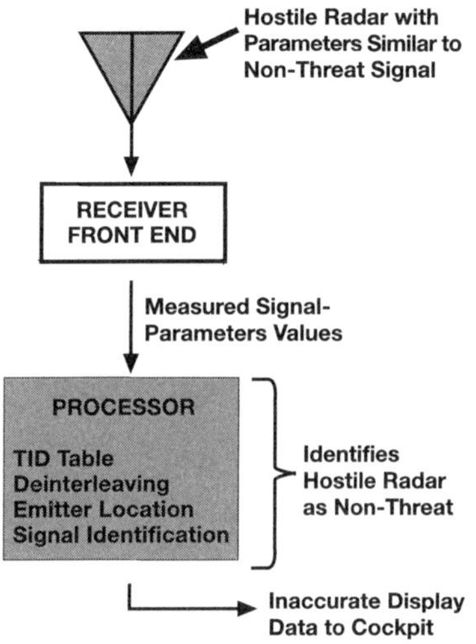

Figure 3.36 A LPID radar has parameters similar to those of a friendly radar or nonthreat signal, thus making it difficult for an ESM receiver to properly identify it.

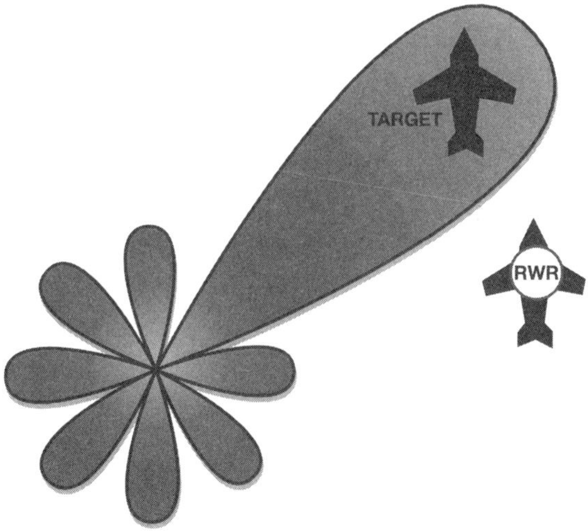

Figure 3.37 A less stringent definition of LPI would be that a radar can detect a target at the same range at which a receiver not in its main beam can detect it.

other (nonthreatening) signals that may be received. Before the processor can identify a signal, it must first isolate that single signal from the many signals present. For pulsed radars, this is called "deinterleaving." For all signals, it involves measurement of frequency, modulation parameters, and direction of arrival along with the sorting of data by parametric values. Once an individual signal is isolated, the processor compares its parameters against the TID table to find a match to a threat or nonthreat signal. Then the ESM receiver reports the presence, operating mode, and location of the identified threat type to cockpit displays.

If a radar uses parameters similar to a friendly type of radar, the ESM receiver is likely to identify it as such, and thus not report the presence of the threat even though it was clearly received. Another approach is to introduce parametric agility. It is far easier for an RWR to identify threat signals with fixed parameters. Agile signals, particularly if that agility causes random parametric changes, require additional analysis time even if the parameters are known.

The shortcomings of the LPID approach are that ESM processing is becoming more sophisticated, and that a radar needs certain information from its modulation to perform its mission. The increasing processor power in modern ESM receivers allows them to more effectively deal with agile parameters and to perform functional and pattern analysis against signals that do not fit the TID choices. More sophisticated processing and precision emitter location techniques will also allow future ESM receivers to perform location correlation and motion analysis to separate friendly signals from imitation friendly signals on hostile platforms.

3.9.4 Detection Versus Detectability

The range at which a radar can detect a target is given by the by the formula:

$$R_{DR} = \text{Antilog}\{[P_T + 2G - 103 - 20\log(F) + 10\log(\sigma) - S_R]/40\}$$

where

R_{DR} = the radar detection range (in kilometers);

P_T = the radar's transmitter output power (in dBm);

G = the radar's peak main-beam antenna gain (in decibels);

F = the operating frequency of the radar (in megahertz);

σ = the target radar cross section (in square meters);

S_R = the sensitivity of the radar's receiver (in dBm).

The range at which a receiver can detect a radar signal is given by the formula:

$$R_{DRCVR} = \text{Antilog}\{[P_T + G_{R/RCVR} - 32 - 20\log(F) + G_{RCVR}]/20\}$$

where

R_{DRCVR} = the receiver's detection range (in kilometers);

P_T = the radar's transmitter output power (in dBm);

$G_{R/RCVR}$ = the radar antenna gain in the direction of the receiver (in decibels);

F = the operating frequency of the radar (in megahertz);

G_{RCVR} = the receiver antenna gain (in decibels);

S_{RCVR} = the sensitivity of the receiver (in dBm).

These equations apply in both Figures 3.37 and 3.38. By selecting some values to plug into these formulas and assigning bandwidth and processing gain values (which drive the sensitivities) we will be able to investigate LPI radar performance in realistic cases.

3.9.5 LPI Figure of Merit

The LPI figure of merit for a radar could be considered the ratio of range at which the radar can detect a target to the range at which its signal can be detected by an ESM receiver.

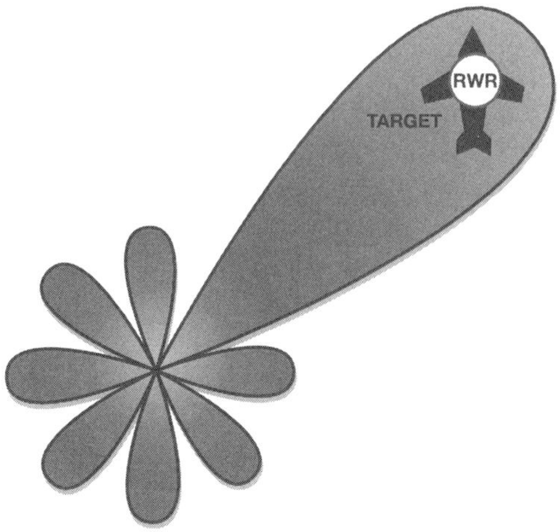

Figure 3.38 The most challenging definition of LPI is the "quiet" radar. It can detect a target at the same range that a receiver on the can detect the radar.

The ratio of the receiver detection range to the radar detection range increases with the receiver's antenna gain and decreases with the radar cross section of the target. It also increases as the ratio between the sensitivity level of the receiver and that of the radar decreases.

To avoid confusion on the sensitivity issue, remember that the sensitivity level is the lowest signal that a receiver can accept and still do its job. Thus as the sensitivity improves, the sensitivity level decreases. The sensitivity numbers in both of the range equations earlier are large negative numbers (sensitivity level)—therefore, as each sensitivity improves, the corresponding detection range increases.

3.9.6 Other Factors Impacting Detection Range

There are two considerations that do not appear in these formulas. One is that the radar's detection range is really controlled not by the peak power, but by the energy that is reflected from the target and coherently processed by the radar. The second is that several factors impact the sensitivity.

3.9.6.1 Coherently Processed Energy from the Target

The radar range equation for received energy also has a factor for time on target, which is the time over which the radar can coherently integrate the return signal. Thus, the radar range can be expressed as a function of average power and time on target. This can be increased either by increasing the power, or increasing the duration of the signal—as long as the target remains in the radar's antenna beam and the return can be coherently integrated.

Another constraint on the radar is that its ability to resolve the range to the target is determined by the pulse width. Range resolution is commonly defined as:

$$\Delta R = \tau c / 2$$

where

ΔR = the range resolution;

τ = the pulse width;

c = the speed of light.

Modulations on the signal allow better range resolution for any given signal duration. This modulation can be frequency modulation (chirp) or phase reversals (binary phase shift keying), as explained in Section 3.5. The radar can also use other phase modulations such as quadrature phase shift keying (QPSK) or higher level phase modulation (M-ary PSK).

Since the detecting receiver depends on the peak power of the radar signal to detect the radar, the radar can gain an advantage in detection range by using a lower power, longer duration signal along with some modulation which will allow it to achieve adequate range resolution (see Figure 3.39).

3.9.6.2 Sensitivity Factors

We ordinarily consider receiver sensitivity to be a function of bandwidth, noise figure and required signal-to-noise ratio. As shown in Figure 3.40, the sensitivity in dBm (i.e., lowest signal the receiver can accept and still perform its function) is the sum of kTB (in dBm), noise figure (in decibels) and required signal-to-

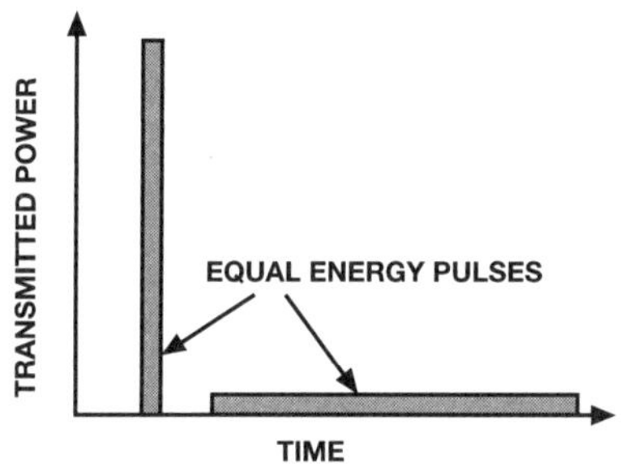

Figure 3.39 By increasing its pulse duration, a radar can reduce its transmitted power level without diminishing the energy it places on the target.

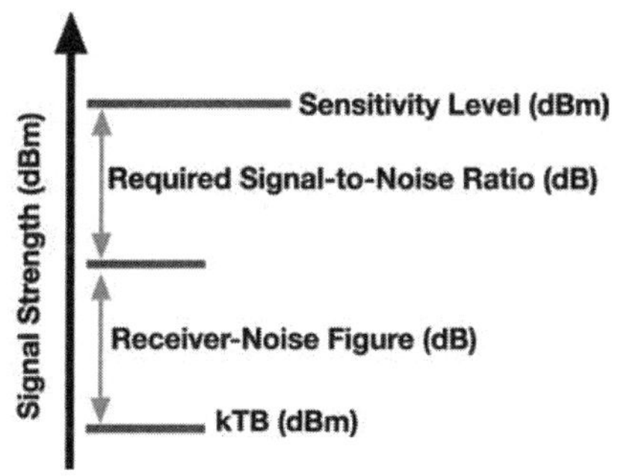

Figure 3.40 Receiver sensitivity (in dBm) is the sum (in decibel terms) of kTB, noise figure, and required signal-to-noise ratio.

noise ratio (in decibels). In radar analysis problems, the required signal-to-noise ratio is usually set to 13 dB and kTB is usually taken as:

$$kTB = -114 \text{ dBm} + 10 \log (BW)$$

where

kTB is the thermal noise in dBm;

BW is the effective receiver bandwidth in megahertz.

However, there is another factor that is useful to consider in the context of LPI signals; that is processing gain. Processing gain has the effect of narrowing the effective bandwidth of the receiver by taking advantage of some aspect of the signal modulation. The advantage comes when the radar's receiver can achieve the processing gain—while a hostile receiver cannot.

A radar achieves bandwidth advantage over an intercept receiver because it can match its receiver and its processing to its own signal—while the intercept receiver must accept a wide range of signals and must typically make detailed parametric measurements to identify the type of signal it is receiving. For example, a pulse radar need only determine the round-trip pulse travel time—and can integrate several pulses to determine that time. It doesn't care about the shape of the pulse it receives, so its effective bandwidth (including processing gain) is significantly less than the inverse of the pulse width. The intercept receiver, on the other hand, must determine the pulse width. This requires a pulse with definable leading and trailing edges—which, in turn, requires a bandwidth of about 2.5 times or more the inverse of the pulse width (see Figure 3.41).

3.9.6.3 Coherent Detection

An EW receiver cannot achieve coherent detection of radar signals, but an LPI radar can do so because the transmitter is typically located with the receiver. When there is randomness in the signal modulation, this becomes even more pronounced. The most extreme example of this effect is the use of true noise to modulate a radar signal. LPI radars using noise modulation are called random signal radars (RSRs). The RSR uses various techniques to correlate the return signal with a delayed sample of the transmitted signal as shown in Figure 3.42. The amount of delay required to peak the correlation determines the range to the target. Since the transmitted signal is completely random and the intercepting receiver has no way to correlate to the transmitted signal, it can only determine that the radar is present through energy detection techniques rather than detecting modulation characteristics. This is a much less efficient process than that available to the radar receiver.

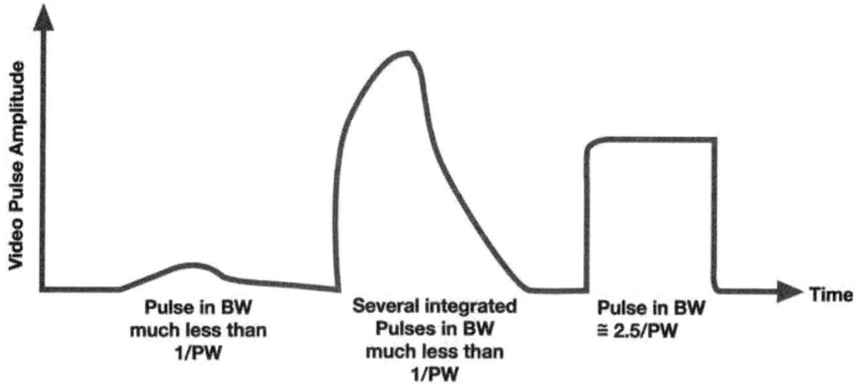

Figure 3.41 If a receiver has bandwidth less than the inverse of the pulse width, the pulse parameters are very difficult to measure, but such pulses can be integrated to determine pulse time of arrival. With bandwidth greater than the inverse of the pulse width, the pulse parameters can be measured.

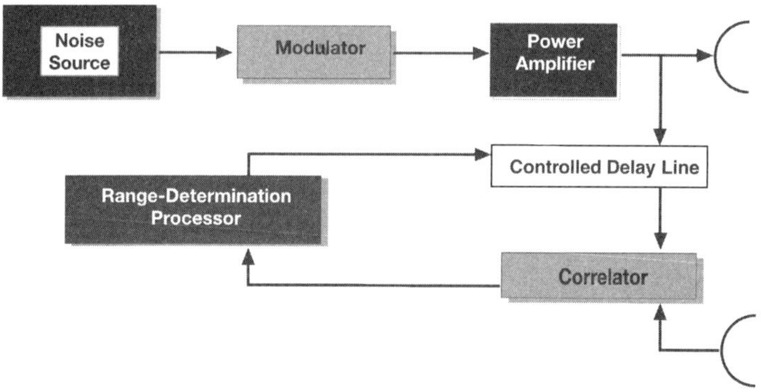

Figure 3.42 A random-signal radar transmits signals with random modulation. It determines the range to the target by correlating return signals with a delayed sample of the transmitted signal.

3.9.6.4 Current LPI Radars

Several radars considered to be LPI have been developed and deployed over the last 20 years. They have used frequency modulation and phase coding to achieve range resolution while reducing their detectability by transmitting long-duration/low-power signals. There are also several such systems under current development, and random signal radars are described in technical literature.

In all cases, the radars' level of LPI is described in terms of range detection ratio with various engagement parameters specified (for example the target cross section). They are also described in terms of "warning time" which is the time between the detection of the radar by a hostile receiver carried by a target and the detection of that target by the radar. Again, engagement parameters must be specified (e.g., target approach speed and radar cross section and the type of receiver employed).

4

Infrared and Electro-Optical Considerations in Electronic Warfare

The object of electronic warfare is to deny an enemy the benefits of the electromagnetic spectrum while preserving those benefits for our own forces. It is easy to fall into the trap of considering only the radio frequency part of the electromagnetic spectrum, but there is a significant amount of EW effort in the IR, visible light, and ultraviolet parts of the electromagnetic spectrum. In this chapter, we will deal with the general nature of these parts of the spectrum, the systems that operate in this range, and the nature of the countermeasures against those systems.

4.1 The Electromagnetic Spectrum

Figure 4.1 shows the portion of the electromagnetic spectrum that is of most interest to the electronic warfare field. Although we typically use frequency to define the RF portions of the spectrum, it is more common to use wavelength at the higher frequencies. Note that wavelength and frequency are related by the speed of light in the formula:

$$c = f\lambda$$

where

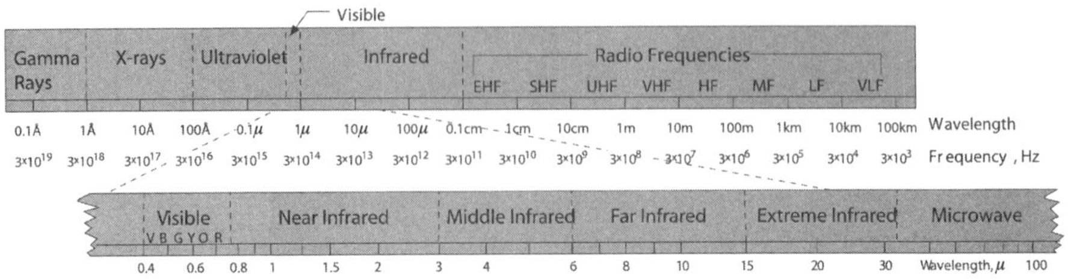

Figure 4.1 The electromagnetic spectrum includes RF, IR, visible, and above-visible frequencies.

c = the speed of light (3×10^8 m/s);

f = frequency (in hertz);

λ = wavelength (in meters).

Frequencies below 300 GHz (i.e., wavelengths longer than 0.1 cm) are in the radio frequency range. Above that frequency, we'll only talk about the wavelengths. The common unit of wavelength is the micron (10^{-6} meter) which is abbreviated μ. At very short wavelengths, angstroms are used (10^{-10} meter, abbreviated Å).

- From about 30μ to about 0.75μ is the infrared region;
- From about 0.75μ to about 0.4μ is the visible light region;
- From about 0.4μ to about 0.01μ is the ultraviolet region;
- Shorter waves than these are X-rays and gamma rays (these regions overlap).

4.1.1 Infrared Spectrum

The infrared region is generally divided into four more definitive ranges:

- The near infrared wavelength range starts at the upper edge of the visible light region (about 0.75μ) and ends at 3μ.
- The middle infrared range is from 3μ to 6μ.
- The far infrared range is from 6μ to 15μ.
- The extreme infrared range has wavelengths greater than 15μ.

In general, hot targets emit most of their IR energy in the near-IR region. This includes the rear aspect view of a jet engine (looking up into the engine). Most of the Sun's IR energy is also in the near-IR range. Slightly cooler targets (such as hot metal parts on the exterior of a jet engine and the jet engine's plume) emit most of their IR energy in the mid-IR region. Objects in the normal range of temperatures (i.e., the skin of aircraft and other vehicles, clouds, and the Earth) emit in the far-IR region.

4.1.2 Blackbody Radiation

A blackbody is a theoretical ideal IR radiator very useful in the study of IR systems and countermeasures. While there are no true "blackbodies," everything that emits IR energy does so in a pattern similar to that of the blackbody model. IR radiation is stated in watts per cm^2 μ^{-1}. The IR emissivity of real-world materials is defined in terms of a percentage of blackbody radiation at a given temperature. Generally, the emissivity values vary from 2% to 98%. Typical examples of emissivity are: polished aluminum 5% at 100°C, average color paint 94% at 100°C, snow 85% at –10°C, and human skin 98% at 32°C.

The blackbody radiation versus wavelength is a function of the temperature of the emitter. As shown in Figure 4.2, there is considerably more energy under the curves for higher temperatures. The total energy varies as the fourth power of temperature. Also, the peaks of the curves move to lower wavelengths as the temperature increases. Figure 4.3 shows a logarithmic curve of radiation

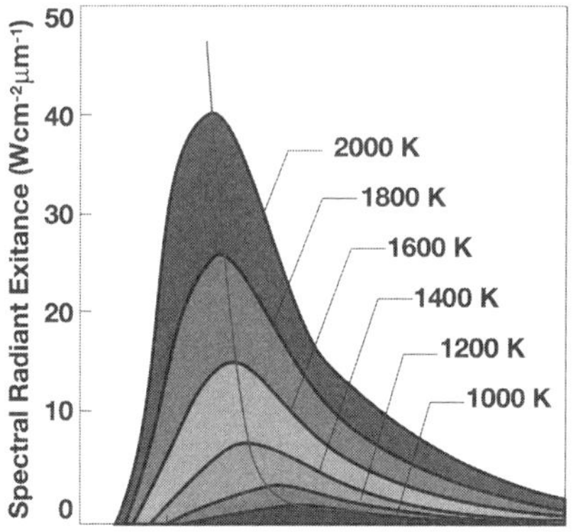

Figure 4.2 Radiation from a blackbody has a distinct radiation-versus-wavelength distribution based on its temperature.

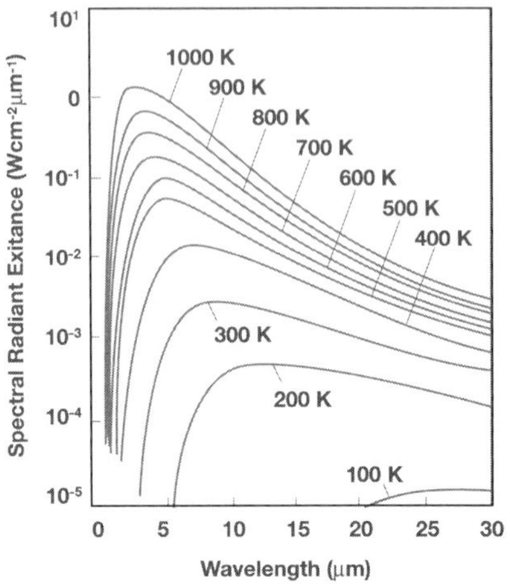

Figure 4.3 The blackbody-radiation curves continue to low temperatures.

versus wavelength for lower temperatures. Note that the curves in the two blackbody figures are for temperatures in degrees kelvin, so 300° is about room temperature. An interesting point is that the surface of the Sun is about 5,900K, which causes its radiation to peak in the visible light spectrum (convenient for those of us with eyes that operate in that range).

4.1.3 IR Transmission

Figure 4.4 shows the relative IR transmission through the atmosphere as a function of wavelength. Note that there are absorption lines from various atmospheric gasses, but there are major transmission widows in the near-, mid- and far-infrared ranges.

In IR transmission, the spreading loss versus range is calculated by projecting the receiving aperture from its range onto a unit sphere around the transmitter as shown in Figure 4.5. The spreading loss is then the ratio of the amount of the surface of the unit sphere covered by the image of the receiving aperture to the whole surface area of the sphere. This is actually the same way we calculate spreading loss for RF signals. However by assuming isotropic antennas we get range and frequency terms in the RF equation.

4.1.4 EW Applications in the IR Range

EW systems and threats receive IR energy to detect, identify, locate, and guide missiles to radiating objects. Examples of these systems and threats are: IR line scanners, forward looking infrared (FLIR), and IR-guided missiles.

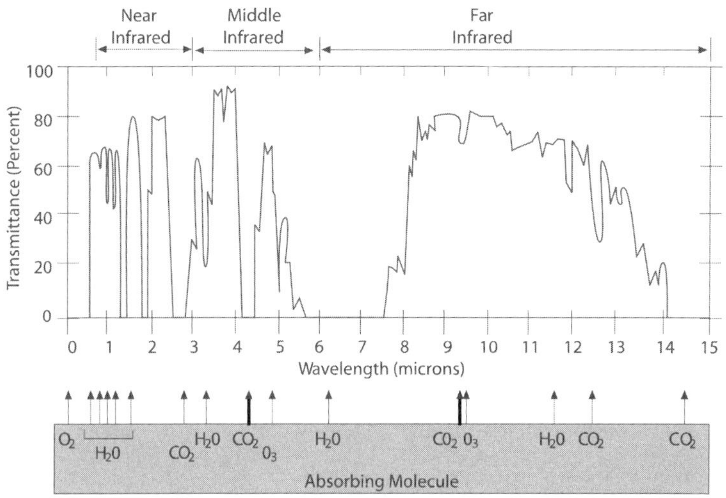

Figure 4.4 This is the percentage transmission through 6,000 feet of atmosphere at sea level with 17 mm of precipitable water.

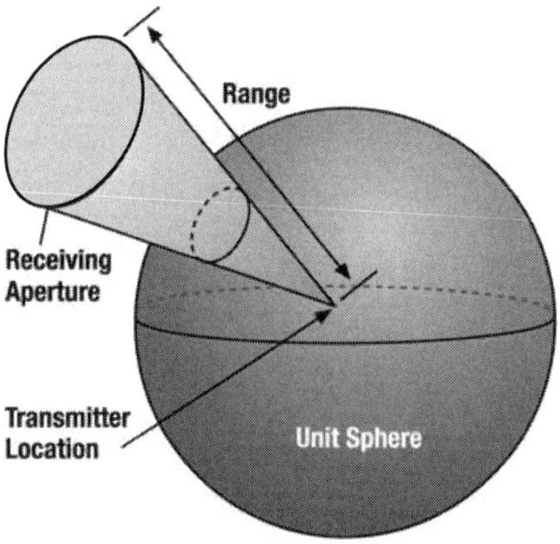

Figure 4.5 The spreading loss for IR transmission is the ratio of the receiving aperture projected on a unit sphere around the transmitter to the surface area of the sphere.

There are, of course, countermeasures to all of these systems. Sensors can be blinded (temporarily or permanently) and IR-guided missiles can be defeated by flares or IR jammers.

4.1.5 Electro-Optical Devices

We are making a somewhat arbitrary distinction between IR and electro-optical (EO) devices to isolate devices receiving radiated IR energy from the rest of the field of interest. Some of these EO devices operate in the infrared spectral range. EO systems and applications (and their countermeasures) discussed in this chapter include:

- Laser communication;
- LADARs;
- Laser range finders;
- Laser designators for missile attack;
- Imagery-guided missiles;
- High-power laser weapons;
- Low-light television;
- Daylight television.

4.2 IR Guided Missiles

IR-guided missiles have been among the most deadly threats in all recent conflict. They are primarily air-to-air missiles and surface-to-air missiles, and include small, shoulder-fired weapons. An IR missile detects the IR signature of an aircraft (against a cold sky) and homes on energy in one of the three IR bands. Early IR missiles required high-temperature targets, so they needed to see the hot internal parts of jet engines to achieve good performance. Therefore, they were usually restricted to attacks from the rear aspect of jet aircraft. More recent missiles can operate effectively against cooler targets (the plume, the tailpipe, heated leading edges of wings, or the IR image of the aircraft itself). This allows them to attack all types of aircraft from all aspect angles.

4.2.1 IR Sensors

Original missiles used uncooled lead sulfide (PbS) detectors that required targets in the 2 to 2.5 micron range (near IR band). This type of missile suffered from considerable solar interference and severely restricted air-to-air tactics.

More modern seekers use sensors of lead selinium (PbSe), mercury cadmium teluride (HgCdTe), and similar materials that operate in the mid- and far-IR bands. While these seekers allow all aspect attack, they require that the sensors be cooled to about 77K with expanding nitrogen.

4.2.2 IR Missiles

Figure 4.6 is a diagram of an IR-guided missile. The nose of the missile is an IR dome. This is a spherical protective covering for the seeker optics, made from a material that has good transmission of IR energy. A seeker senses the angular location of the IR source and hands-off error signals to the guidance control group, which steers the missile toward the target by control commands to the rollerons.

Figure 4.7 is a diagram showing the functions of a simple IR seeker (in a cross section). There are two mirrors (a primary and a secondary reflector) that are symmetrical around the optical axis. They focus energy through a reticle onto an IR sensing cell. Not shown in the diagram is the filter that limits the spectrum of signals passed through the reticle—and the sensor cooling if required.

A simple, spinning reticle pattern is shown in Figure 4.8. Often called the "rising sun" pattern, it rotates around the optical axis of the seeker. The top half of this reticle is divided between very low and very high transmission segments. An IR target is shown in one of the high transmission segments. The other half of the reticle has 50% transmission. This reduces the dynamic range required of the IR sensor. As the reticle rotates, the IR energy from the target onto the IR sensor will vary in the partial square wave pattern shown in Figure 4.9. The square wave portion of the waveform starts as the upper half of the reticle starts to pass the target. Since the sensor knows the angular position of the reticle, it can sense the direction to the target from the timing of the square wave portion

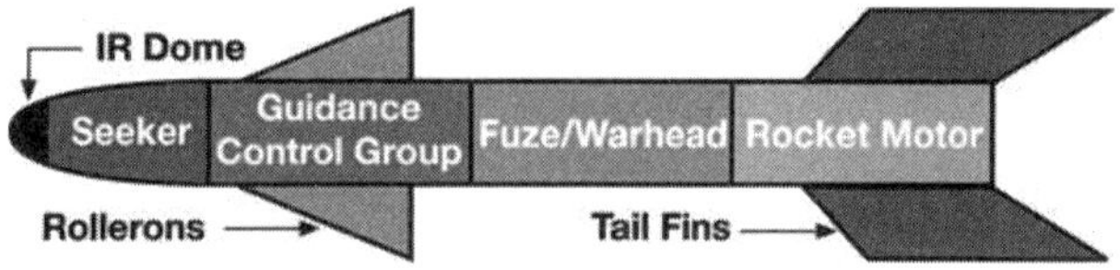

Figure 4.6 A heat-seeking missile is guided by input from an IR sensor.

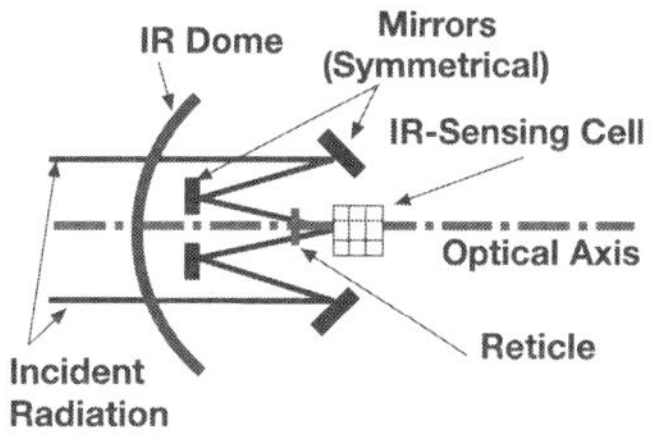

Figure 4.7 The IR seeker focuses received IR radiation on a sensing cell through a reticule.

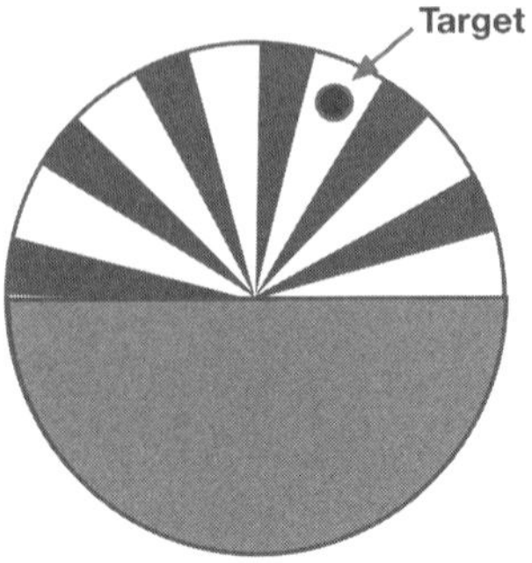

Figure 4.8 The reticule modulates energy from a target heat source as a function of its position relative to the seeker.

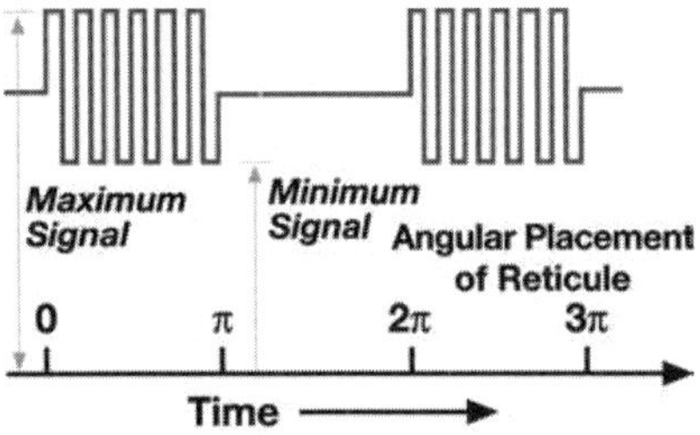

Figure 4.9 The rotating reticule creates a pattern that allows determination of the correction direction toward the heat source.

of the waveform. This allows the guidance group to make corrections to steer the missile toward the target.

Figure 4.10 shows the amount of maximum signal power as a function of the angular offset of the target from the optical axis of the sensor. When the target is near the center, the high-transmission segment does not admit the whole

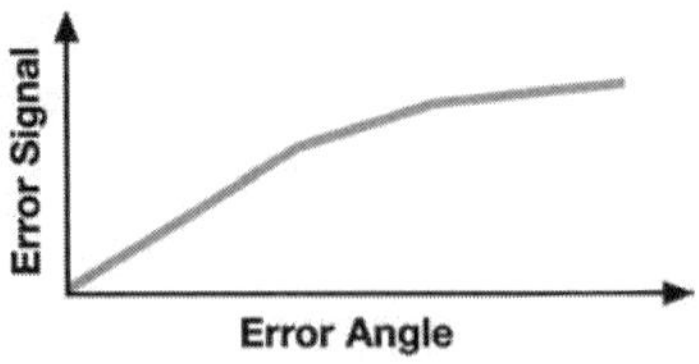

Figure 4.10 The error signal from the reticule in Figure 4.8 flattens as soon as the high-transmission segment is wide enough to pass the whole target.

target. As it moves farther from the center, more of the target is passed. Once the whole target is passed, the peak energy level to the IR sensor does not increase more. This means that the sensor only provides proportional correction inputs when the target is quite near the center of the reticle. It also means that the seeker has no way to discriminate against a high-energy false target near the outer edge of the reticle.

Figure 4.11 shows a nonrotating reticle with a "wagon-wheel" pattern. To generate steering information, the energy entering the seeker is nutated to move it around the optical axis. If the target is on the optical axis, it will cause a constant amplitude square wave of energy to reach the sensor. However, if the target is off center, its image will move in the offset circle shown in the figure. This causes the irregular square wave form shown at the bottom of the figure. The control group can then determine that the missile must steer in the direction away from the narrow pulses.

Figures 4.12 and 4.13 show two of the many types of more complex reticles. Figure 4.12 shows a multiple frequency spinning reticle. Since there are different numbers of segments in each of several rings, the number of pulses seen by the sensor changes as a function of the angular distance from the optical axis to the target. This supports proportional steering. Figure 4.13 shows a spinning reticle with curved spokes to discriminate against straight-line objects (like the horizon) and has a different numbers of spokes at different offset angles for proportional steering.

To avoid extremely high g forces on a missile as it reaches its target, missiles use proportional navigation as shown in Figure 4.14. If the aircraft and the

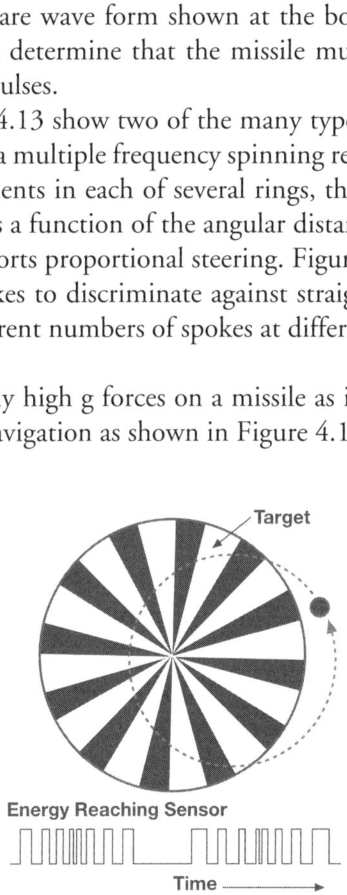

Figure 4.11 The "wagon-wheel" reticule remains fixed, while the image of a fixed target is nutated, causing an irregular pulse pattern when the target is away from the optical axis.

Figure 4.12 The multiple-frequency reticle has different numbers of segments at different distances from the center of rotation.

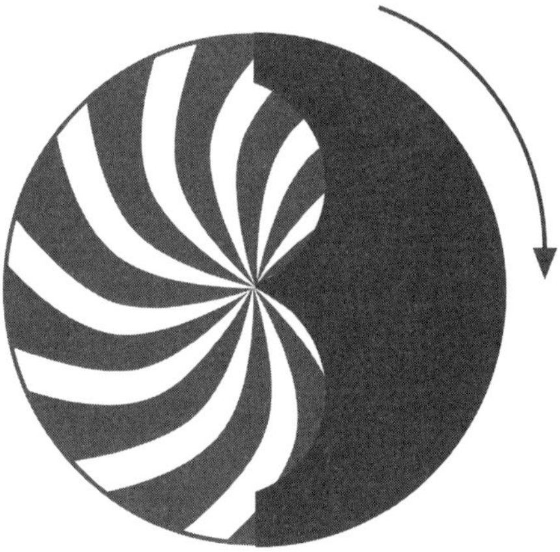

Figure 4.13 The curved spokes discriminate against straight-line objects.

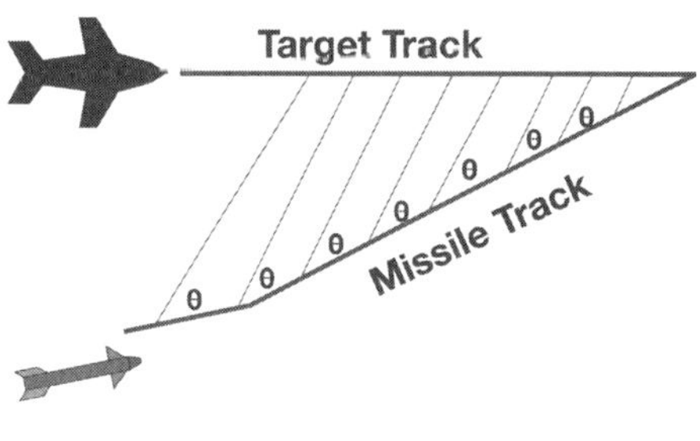

Figure 4.14 Proportional navigation allows the missile to intercept the target with minimum g forces.

missile are both at constant velocity, a constant offset angle (θ) between the missile's velocity vector and the optical axis of its seeker will cause an optimum intercept. If either is accelerating (e.g., the target is taking evasive action) corrections must be made to return the missile to the proper offset angle.

4.3 IR Line Scanners

The infrared line scanner (IRLS) is one of several IR devices that are useful in various types of reconnaissance applications. The IRLS provides an IR map of a covered area. It is mounted on a manned aircraft or unmanned aerial vehicle (UAV) which flies a fairly low-level path over an area of interest. The IRLS makes a two-dimensional image by scanning an IR detector over an angular increment across the ground track of the vehicle while the second dimension is provided by the movement of the platform along its ground track.

4.3.1 Mine Detection Application

There are a number of military and civil IRLS applications, but the nature and limitations of the IRLS can be well understood from its use in the detection and location of buried mines. This approach to mine detection is practical because buried mines will gain or lose heat at a different rate than the surrounding soil (or sand). Thus, the mines will be at a different temperature during times of temperature change, for example, right after sunset. However, the resolution of the IR sensor must be adequate to differentiate the temperature of the mine from that of the soil, and it must have adequate angular resolution to differentiate mines from other buried objects, for example, rocks.

4.3.1.1 An Example

Let's assume that a buried mine is about 6 inches in diameter, but that the sensor must have a resolution capability of 3 inches to identify mines with adequate accuracy. Furthermore, let's assume that the aircraft or UAV is flying at 100 nautical miles per hour (knots), and that the IRST scans a 60° segment across the ground trace as shown in Figure 4.15. Finally, let's assume that the IR sensor has a 0.25-mrad aperture and that the IR energy level is digitized with 8 bits. This high resolution will be required because the soil can have a relatively wide temperature range and post mission analysis will probably be required to find the relatively narrow temperature difference between the mine and the soil anywhere in this range.

First, let's determine how high the aircraft can fly and still get the necessary 3-inch resolution from its sensor. The altitude required is:

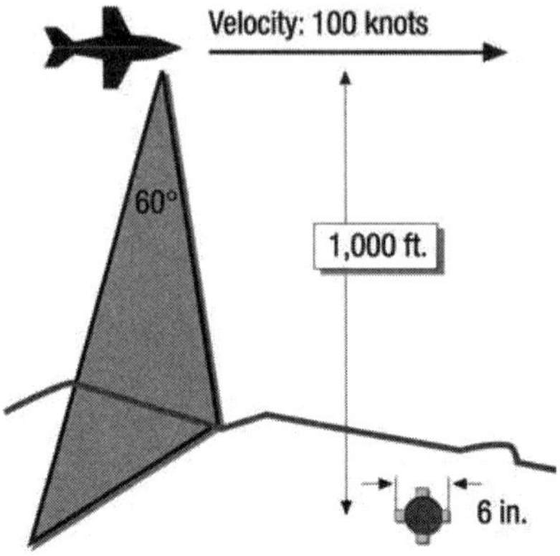

Figure 4.15 A manned aircraft or UAV is here searching a swath along its ground trace with an IR sensor to detect mines as small as 6 inches. It is traveling at 100 knots, 1,000 feet above the ground.

Ground resolution distance versus sin(sensor aperture angle)

At a 0.25-mrad aperture angle, we achieve a 3-inch resolution at 1,000 feet. Figure 4.16 shows the ground resolution versus altitude for a quarter-milliradian sensor instantaneous field of view.

The vehicle can fly at any speed, but the sweep rate must be fast enough to make one cross-track sweep every 3 inches along the flight path. At our chosen speed, 100 knots, the vehicle travels over the ground at 169 feet per second:

$$100 \times (6{,}076 \text{ ft/hr}) / (3{,}600 \text{ sec/hr}) = 169 \text{ ft/sec}$$

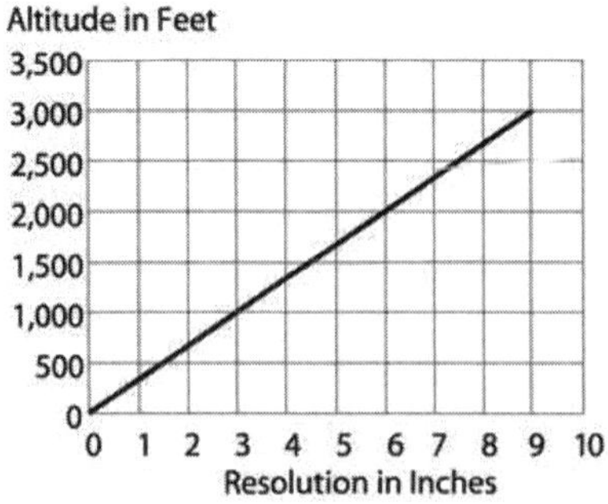

Figure 4.16 The ground resolution provided by a 0.25-milliradian sensor is shown as a function of altitude above the searched terrain.

One sweep per 3 inches requires 4 sweeps per foot or 676 sweeps per second at 100 knots.

The sampling of the IR sensor must also be performed for every 3 inches of movement of the sensor over the ground in the crosstrack sweep. The width of the cross-track ground coverage (the swath width) is:

$$2 \times \sin\left(1/2 \times \text{scan angle}\right) \times \text{altitude} = 2\sin(30°) \times 1{,}000 \text{ ft} = 1{,}000 \text{ ft}$$

A sample each 3 inches requires 4,000 samples per scan. The sample rate is then 676 × 4,000 = 2.7 million samples per second. At 8 bits per sample, this produces a data rate of 21.6 Mbps. With a typical 16% data overhead, this becomes 25 Mbps.

The ratio of velocity to altitude (V/H) is an operational parameter that effects the required data rate. It is stated in radians per second. To understand this choice of units, consider observing the aircraft from a fixed point below it on the ground. Remember that a radian is the angle observed from the center of a circle for one radius along the circumference of a circle. Thus the subtended angle-per-unit time, converted to radians, would be equal to the velocity divided by the radius (i.e., the altitude over the ground). Figure 4.17 shows the altitude to velocity relationship for a V/H of 0.174 rad/sec (which holds the ground resolution distance at 3 inches and minimizes the altitude at the specified data rate for a particular mine resolution payload).

Figure 4.18 shows the data rate required to recover mine resolution data over a 60° swath width as a function of altitude and vehicle speed. As you can see from this typical example, the detection of buried mines requires an airborne platform that flies low and slow, and the collection and analysis of a great deal of data. Detection of larger objects (for example, tanks in underground bunkers) would require less angular resolution. This will allow operation at higher

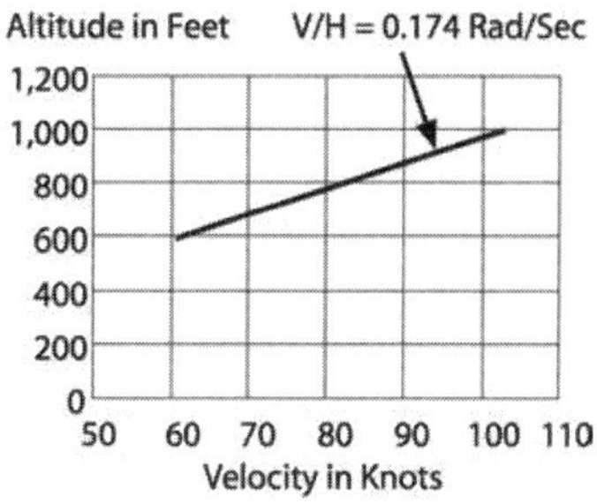

Figure 4.17 The altitude-to-velocity relationship is shown meeting the V/H specification (0.175 rad/sec) for a specific mine-detection payload.

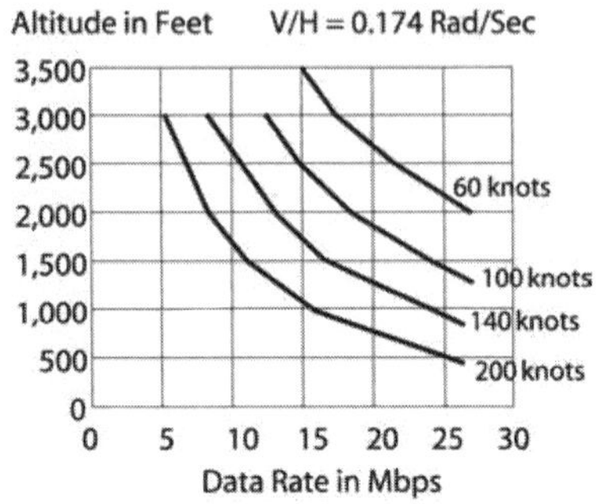

Figure 4.18 The data rate required to cover a 60° swath with 3-inch resolution is shown as a function of altitude.

altitudes and speeds and/or large swath widths. However, a high-data rate should always be expected for IRLS applications because of the fine temperature resolution and large temperature range required. If the vehicle is unmanned, or for any other reason the data is linked to a ground station, a wide data link will be required.

4.4 Infrared Imagery

Imagery involves the capture and display of a two-dimensional picture. This can be in the visible light wavelength range (television) or it can be in a nonvisible wavelength range. Our concern here is imagery at infrared wavelengths. For all electronically implemented imagery, the displayed picture is divided into pixels. A pixel is a spot on the screen; there must be enough pixels to create the picture to the required quality. The system captures and stores the brightness or the brightness and color to be displayed in each pixel—then displays the appropriate values at each pixel location on the screen. The screen display can be generated with a raster scan as shown in Figure 4.19, or by an array as shown in Figure 4.20.

If an imagery system is mapping the ground, the relationship between the pixels and the resolvable distance on the ground would be as shown in Figure 4.21. If the system is looking level or up, the same relationship applies, but the resolvable distance is a function of the range to the individual objects being observed.

4.4.1 The FLIR

A forward-looking infrared (FLIR) system captures and displays a two-dimensional temperature field. It operates in the far-infrared region, where

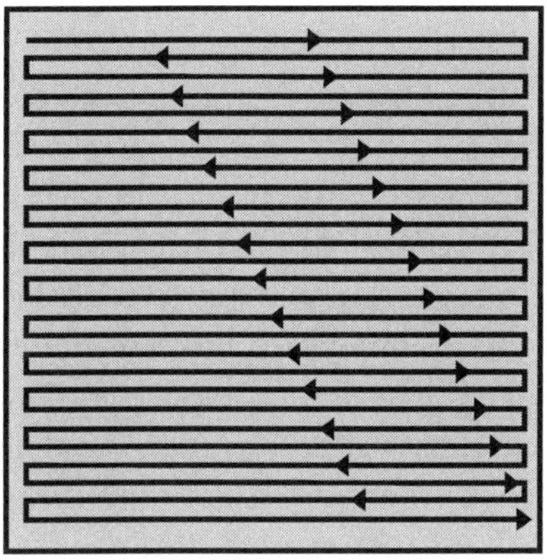

Figure 4.19 A raster scan covers a two-dimensional field. The spacing of pixels on each line is approximately the same as the line spacing.

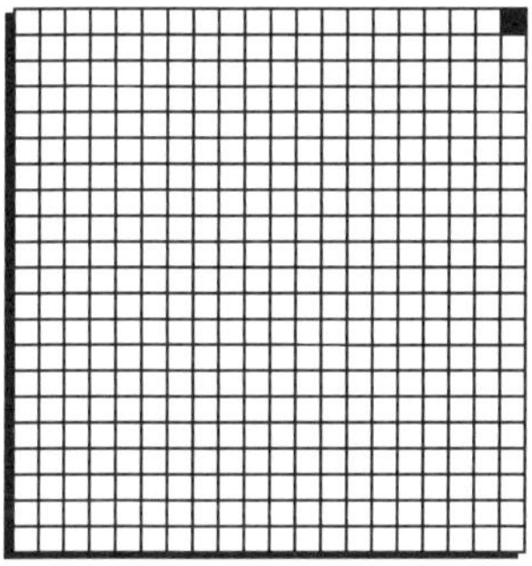

Figure 4.20 A picture can be formed from a group of display points, for example liquid-crystal-display segments. Each element provides 1 pixel.

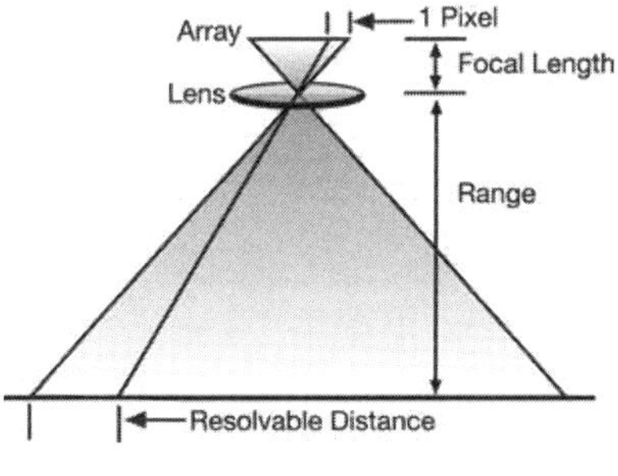

Figure 4.21 Each pixel in an imagery display (or raster) represents one resolvable distance at the range of the object being observed.

everything emits infrared energy. By differentiating the temperatures of objects and backgrounds, the FLIR allows an operator to detect and identify most common objects. The display is monochromatic, with the brightness level of each pixel indicating the temperature at that position in the observed field. FLIRs have some advantages over visible-light TV systems in that they can operate day or night. Also, because they differentiate between objects by temperature or IR emissivity, they can often see objects of military significance that are hidden from visible-light TV by foliage or camouflage.

FLIRs can use serial or parallel processing, or two-dimensional IR arrays as shown in Figure 4.22. With serial processing, the FLIR uses mirrors to scan the orientation of a single IR sensor across a two-dimensional field of view in a raster scan. The whole scene is presented on a CRT. Pixels are defined by the number of samples in a scanned line and the spacing between parallel lines. With parallel processing, a row of detectors is scanned through an angular segment to provide two-dimensional area coverage. Each element of the sensor array takes a series of measurements, so the pixels are defined by the number of elements and the number of samples in a scanned line (by each sensor). A two-dimensional array

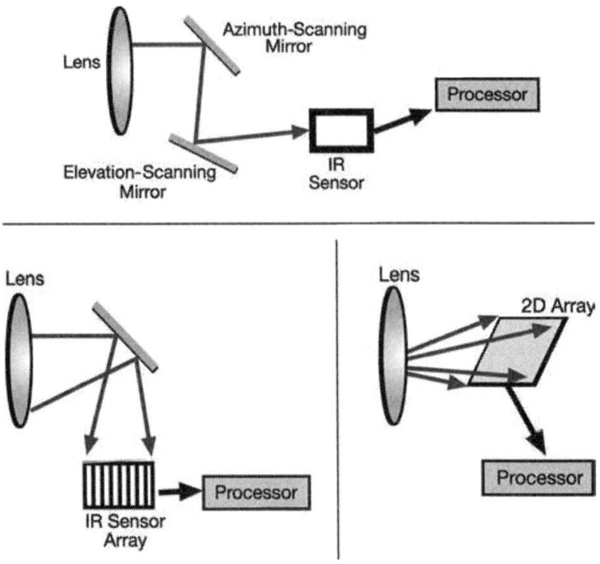

Figure 4.22 (a) A serial-processing FLIR scans a scene with two mirrors to sequentially focus a raster pattern onto a single IR sensor. (b) A FLIR using parallel processing scans a linear array across a scene with a rotating mirror to generate a series of pixels in each sensor. (c) A FLIR using a two-dimensional array of IR sensors instantaneously captures the whole scene being viewed.

captures all of the covered area at once, with each pixel being captured by an array element.

The data rate produced by a FLIR is the product of the number of pixels in a frame (the two-dimensional angular area covered), the number of frames per second, and the number of bits of resolution per sample. Note that one sample is made per pixel.

4.4.2 IR Imagery Tracking

Some modern surface-to-air missiles use imagery guidance. In this approach, the area near the target is observed by a two-dimensional IR array operating in the far-IR region. You will recall that objects at moderate temperatures emit in this region, so the array can observe the contrast between the warmer aircraft and the colder sky. A processor will observe the shape of a number of pixels from the array that show the proper contrast (see Figure 4.23). It will then determine that this pixel distribution qualifies as the target and steer the missile in the corresponding direction. Only a few pixels are required to determine the general size and shape of a target and to differentiate it from a much smaller decoy (as opposed to the large number of pixels required to give a high-quality picture).

4.4.3 Infrared Search and Track

Infrared search and track (IRST) devices are used on aircraft and ships to detect hostile aircraft. An IRST does not use imagery, but rather looks for a warmer spot target against a cold background. It sweeps a large angular area with an IR sensor array as shown in Figure 4.24. It detects IR targets while rapidly covering its angular range. Then, it develops the necessary data to hand off target tracking information to sensors.

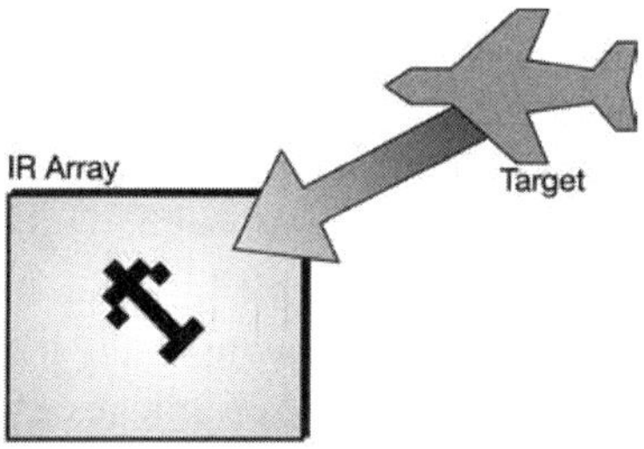

Figure 4.23 An imaging guidance system differentiates the target it is tracking based on a few pixels.

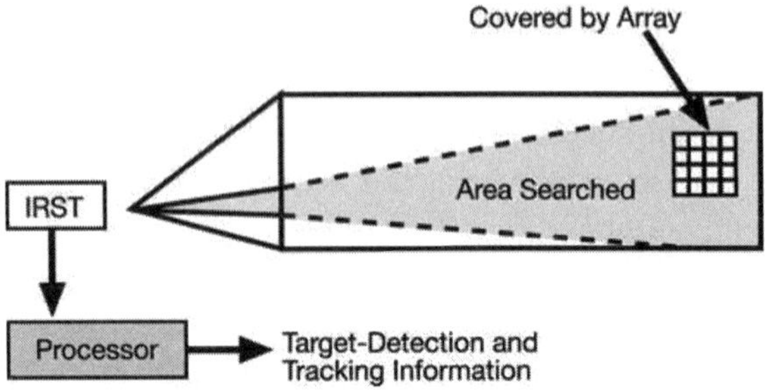

Figure 4.24 An IRST sensor scans a wide angular area with a small array to detect and track warmer spot targets against a cold background.

4.5 Night-Vision Devices

Operation Desert Storm started on the night of January 15, 1991. The reasoning behind the decision to start the operation on that particular date no doubt included complex political and military considerations. However, when you consider that the dark of the Moon occurred on this date, it is obvious that one important factor was the ability of the coalition forces to operate in total darkness while the Iraqis could only bring their full military capability to bear during daylight.

The coalition forces had significant numbers of night-vision devices deployed through their forces and had adequately trained them in their tactical application. These night-vision devices were the product of three generations of development and were completely passive, amplifying very low levels of available light—even on moonless nights with cloud cover.

4.5.1 Types of Devices

Night-vision devices include low-light-level television (L^3TV), viewers for drivers of trucks and tanks, weapon sights, and night-vision goggles for aircrew and ground forces. These devices are different from FLIR devices in that FLIRs receive infrared energy emitted by objects—while night-vision devices amplify available light reflected from objects. The FLIR can operate in total darkness, while night-vision devices require some (though very little) available light.

Light amplification devices have the advantage of being less expensive than FLIRs, and thus available for much wider distribution. Also, since they operate in the optical region, they present the clues necessary for maneuvering of aircraft and ground vehicles and for the movement of troops over terrain. However, since night-vision devices provide no peripheral vision, they require significant training for effective tactical use.

4.5.2 Classical Night Operations

Night operations have always been a part of military action, but they have depended on stealth and the extension of the senses of personnel. For example, consider one of every infantry basic trainee's least favorite training exercises—the infantry platoon in the night attack. The procedure was to maneuver as close to the enemy as possible through the dark. Troops moved single file, each soldier following the luminous strips stapled to the back of the hat of the individual ahead of him in line. The night vision of the troops was carefully guarded by using only red lights to read maps. Troops were trained to keep their eyes moving and to use their peripheral vision (which is more sensitive to light). If you stared directly at an object in the dark (as required to fire a rifle), it would fade out of your vision. Ideally, the troops could sneak close enough to the enemy to move into a final assault line (troops abreast facing the enemy) before being detected. Then artillery-fired flares would light up the battlefield allowing daylight tactics to be used (and completely destroying the night vision of everyone involved).

With modern night-vision devices, troops can move fast and fire with complete accuracy in complete darkness.

4.5.3 History of Development

Before the development of light amplification devices, there were so-called "sniper scopes" which used IR spotlights and IR sensing scopes to fire weapons in "total darkness" (meaning that there was no light that could be sensed by the naked eye). Troops were warned to turn the scope on before the spotlight so they could see any spot lights that were on—and shoot the unlucky enemy who had his light on. You can see a major disadvantage to those devices—which were also used on tactical vehicles.

During the Vietnam War, first generation light amplification devices (called starlight scopes) were used. These could provide a few hundred yards visibility, but emitted an audible "whine" and would "bloom" blanking the entire image when a bright light source was present.

Second generation technology (in the 1980s) included helmet-mounted goggles for helicopter pilots along with sights for rifles and crew-served weapons. These devices provided increased range and rapid recovery from light saturation. However they had short tube life and were subject to saturation by cockpit lighting. Blue/green instrument lighting and a corresponding filter in the goggles were required to prevent this saturation.

Third generation technology provides increased sensitivity, reduced size, improved tube life, reduced blooming, and visibility extension into the near-infrared region. The IR capability allows night vision goggles to see 1.06-micron laser designators.

4.5.4 Spectral Response

Figure 4.25 shows the relative response (versus wavelength) of the human eye as compared to the response of second and third generation light amplification devices.

4.5.5 Implementation

Figure 4.26 shows the operating principle of first generation light amplification devices. Light falling on specially coated electrode screens (dynodes) causes

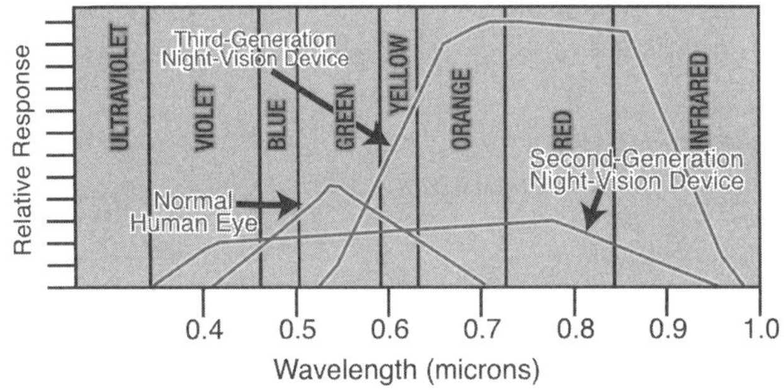

Figure 4.25 Third generation night vision devices operate in both the visible and the IR ranges.

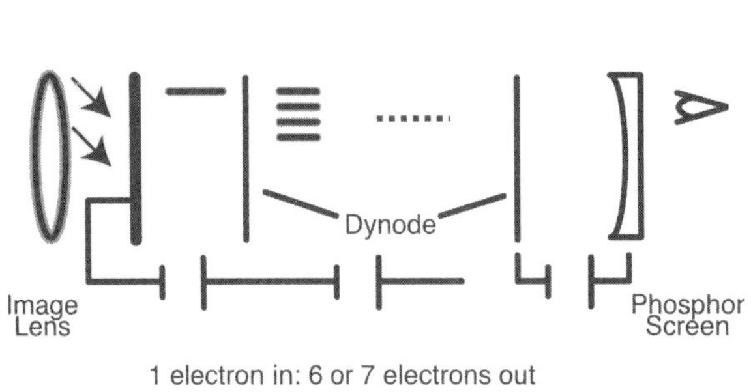

Figure 4.26 First generation light amplification devices amplify in dynodes.

electron emissions which are accelerated in a vacuum by high voltage and kept focused by magnetic fields. These accelerated electrons are converted back to optical images by impact on a phosphor screen. Three stages were required to achieve the required amplification.

Figure 4.27 shows the operating principle of second generation devices. They use a combination of vacuum devices and microchannel plates to achieved the necessary gain.

Microchannel plates are pieces of glass with the order of 10^6 lead-lined holes. Electrons impact the walls of the tubes and dislodge secondary electrons. The secondary emission causes approximately 3×10^4 exit electrons for each primary electron. These secondary electrons are accelerated and focused on a phosphor screen for viewing.

Third generation devices achieve all amplification in microchannel plates as shown in Figure 4.28. The tubes are angled to assure impact of the primary electrons with the lead lining of the tubes.

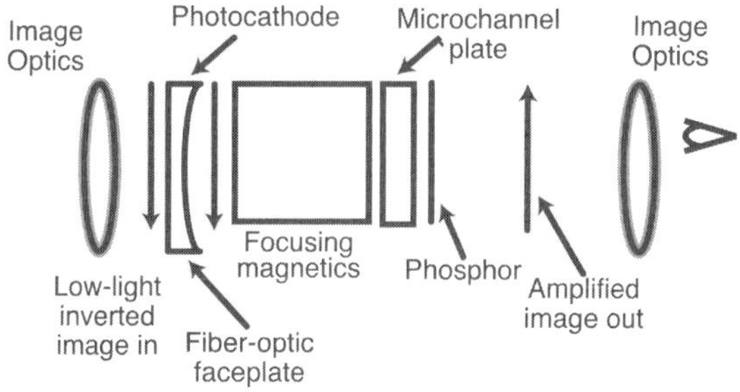

Figure 4.27 Second generation night vision devices combine vacuum and microchannel techniques.

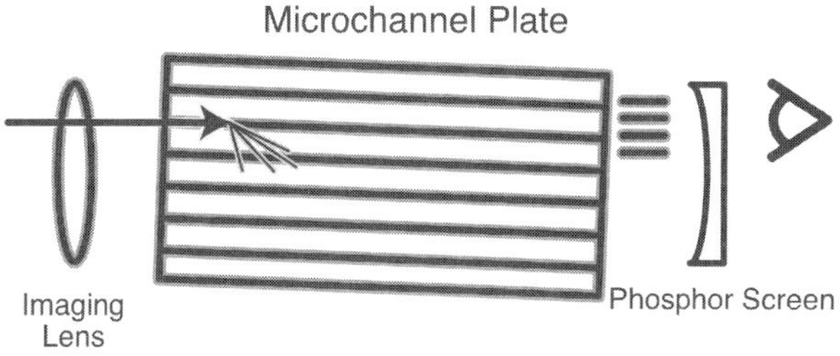

Figure 4.28 Third generation night vision devices generate all gain in a microchannel plate.

4.6 Laser Target Designation

Laser designators and range finders have long been used against fixed and mobile ground targets, and now are also significant threats to helicopters and fixed-wing aircraft.

4.6.1 Laser Designator Operation

When a laser illuminates a target, there is significant energy in the scintillation of laser energy from the surface of the target. A missile with a laser receiver can home on this scintillation, allowing extremely accurate target engagement. Normally, the laser illuminator (called a designator) is coded, improving the ability of the receiver in the missile to discriminate against sun glint and other interfering sources of energy.

The missile must have some sort of guidance scheme (multiple sensors, moving reticule, and so forth) to provide angular error signals for guidance to the target. Its receiver is designed to accept only laser energy at the wavelength of the designator. Its processing circuitry qualifies on the proper coding and converts the angular error signals into guidance commands.

As shown in Figure 4.29, the designator need not be located on the attacking platform. In this case, one aircraft or UAV places the designator on the target and tracks the target if it is moving. A second aircraft fires a missile which homes on the scintillation of the designator from the target. The missiles are fire-and-forget weapons, allowing the attacking aircraft to engage multiple targets and leave the area as soon as the missiles are launched. The designating aircraft must remain within line of sight of the target to keep the designator on the target until missile impact.

Figure 4.30 shows a ground-to-ground engagement with laser designation. The attacking platform places a laser designator on the target and fires its own laser homing missile. Line of sight must be maintained during the whole

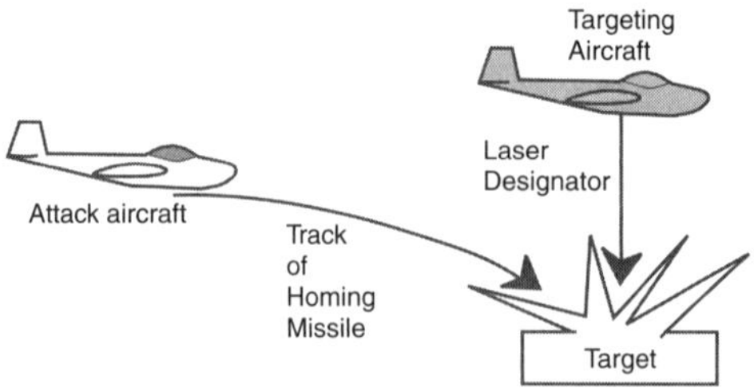

Figure 4.29 A laser designator placed on a target by one aerial platform allows a missile from a second platform to home on the scintillation from the target.

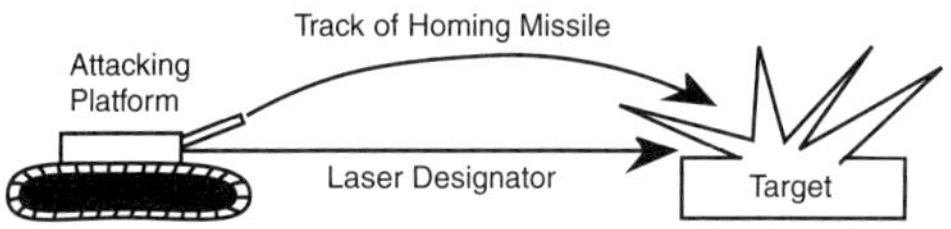

Figure 4.30 A ground mobile weapon can laser designate and fire a homing missile.

engagement to keep the designator on the target. However, in some systems, there is a laser on the missile itself. This allows the attack to continue even if the target maneuvers to avoid line of sight to the attacking platform. Note that a laser range finder on the attacking platform can determine an extremely accurate range to the target, providing a very tight firing solution—which complicates the task of any countermeasures approach.

4.6.2 Laser Warning

The first step in defending against laser-designated weapons is to determine that a laser designator has been placed on the targeted asset. This involves the use of laser detection systems as shown in Figure 4.31, which are used on ground mobile and airborne platforms. These systems typically have four or six sensors distributed around the platform. Since each sensor covers 90° from boresight, four sensors will provide 360° azimuth coverage and about ± 45° in elevation. This is normally adequate for ground vehicles. Aircraft typically have six sensors, providing roughly spherical (4 steradian) coverage.

Each sensor has a lens which focuses incoming laser signals on a two-dimensional array which locates the direction to the laser to a single pixel (1-pixel-per-array element). Greater location accuracy can be accomplished if multiple sensor outputs are used as elements of an interferometer.

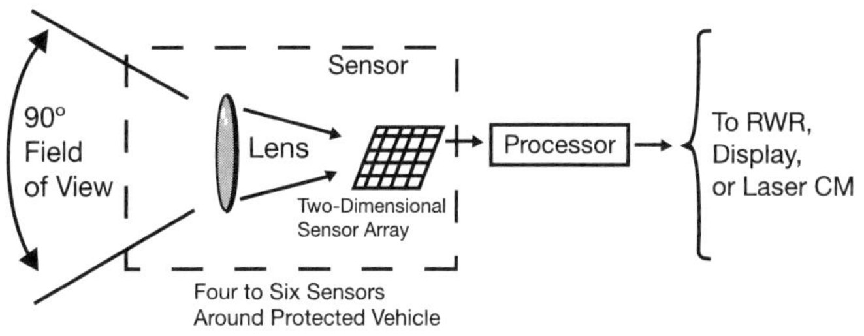

Figure 4.31 A laser warning receiver detects the presence of a laser designator, the type of laser, and the direction of the emitter.

The laser-warning receiver processor determines the type of laser received and the direction of arrival. It either passes this information to a radar-warning receiver (which provides an integrated threat display) or drives its own unique threat display. The laser-warning receiver can also support countermeasures against the laser designator or the associated weapons.

If a low-power laser is scanned through the angular space containing the missile (as in Figure 4.32), the laser will pass through the missile receiver's lens, reflect from the detection array in the missile, and be further enhanced as the reflection passes back through the missile lens to a receiver on the defended asset. By receiving and performing direction of arrival analysis on the reflected signal, the countermeasures system determines the angular location of the missile.

A plume detector is another way that the countermeasure system can locate the missile.

4.6.3 Countermeasures Against Laser-Homing Missiles

There are both active and passive countermeasures against laser-homing missiles.

Active countermeasures (see Figure 4.33) include counterfiring against either the missile or the designator. Since the missile location can be determined by either detecting its plume or detection of laser reflection from its homing receiver, a firing solution for a countermissile missile can be generated. The designator must remain within line of sight of the target, so an accurate laser-warning receiver will give a firing solution for a missile to attack the designating platform (ground or air).

The missile can also be attacked electronically by use of a high-power laser which can either dazzle the missile receiver (saturating its sensor) or actually damaging the sensor.

If a lower power laser has a deceptive jamming signal which causes the missile receiver to pass incorrect error signals to the missile guidance, the missile can be caused to miss its target.

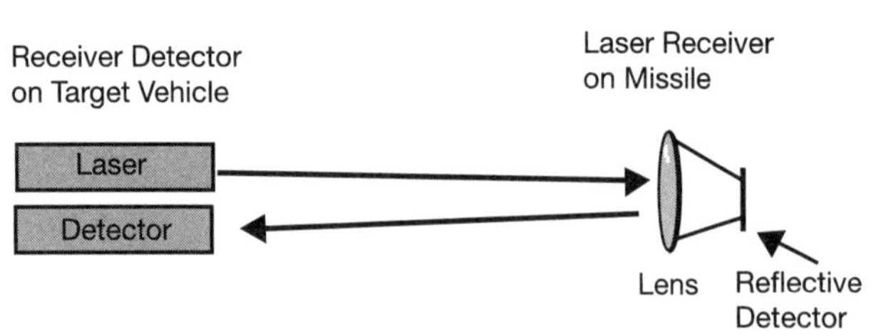

Figure 4.32 The reflection of a laser from the detector in a hostile receiver is enhanced by passing through the hostile receiver's lens twice.

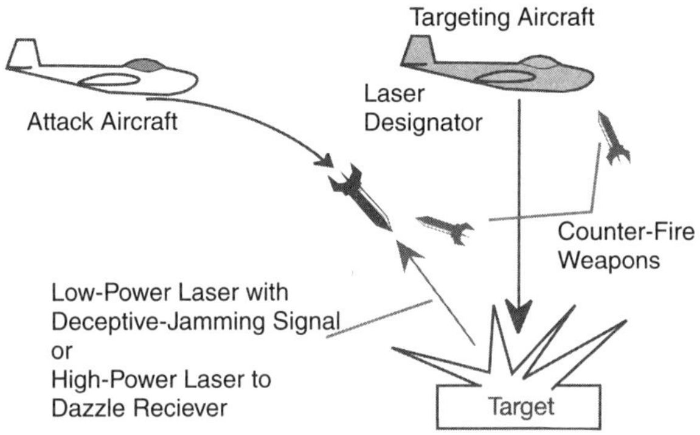

Figure 4.33 Active laser countermeasures include jamming, counterfiring against the designator or the missile, or dazzling the missile receiver.

Passive countermeasures obscure the target, making it difficult for the targeting platform to track the target and keep the laser properly aimed. Obscuration also reduces the power of the scintillation from the laser if it is on the target. Finally, obscuration reduces the propagation of the laser signal to the receiver in the missile, denying it the necessary error signals for guidance.

Smoke formulated to obscure IR, visible light, or ultraviolet signals is an important countermeasure. Water-dispensing systems on ground targets can also generate dense fog around the protected platform—which effectively obscures a wide range of signal frequencies.

4.7 Infrared Countermeasures

Countermeasures to infrared-guided missiles include flares, jammers, decoys, and IR chaff.

4.7.1 Flares

The principal countermeasure against an IR-guided missile has been a high-temperature flare ejected from an aircraft to break the lock of current types of missiles. The flare breaks the lock of the missile onto the aircraft and causes the missile to home on the flare. Although the flare is much smaller than the aircraft it is protecting, it is significantly hotter. Therefore it radiates significantly more IR energy. As shown in Figure 4.34, the missile tracker tracks the centroid of all of the IR energy within its field of view. Since the flare has more energy, the energy centroid is closer to the flare. As the flare separates from the

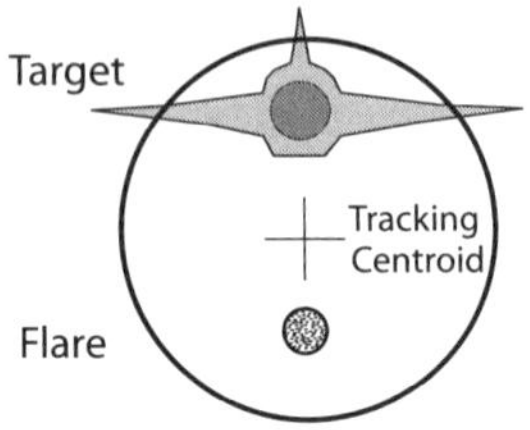

Figure 4.34 A flare has more IR energy than the target, and the missile steers toward the centroid of IR energy in its tracker. Thus, the missile is lured away from the target.

protected aircraft, the centroid is pulled away. Once the aircraft leaves the tracking field of view of the missile, the missile homes on just the flare.

Newer weapons use so-called "two color" trackers to overcome the energy advantage of hot flares. The black body curves in Figure 4.2 show that there is a unique energy versus wavelength curve for each target temperature. As shown in Figure 4.35, the spectral radiance of the flare (at 2,000K) versus wavelength can have a significantly different shape than that of the tracked aircraft, which has a

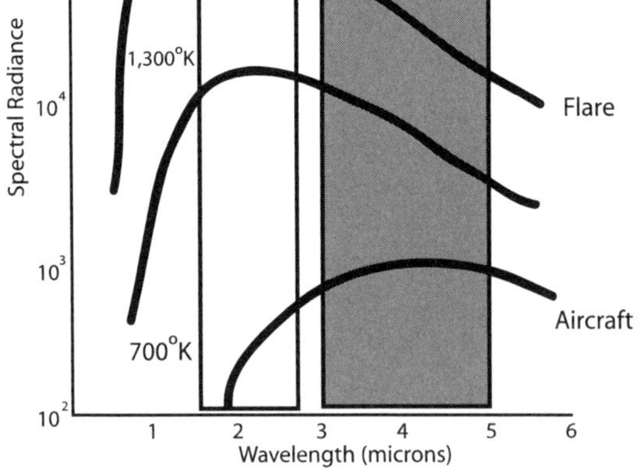

Figure 4.35 A two-color sensor can determine the temperature of its target by comparing the energy at two frequencies.

much lower temperature. By measuring and comparing energy at two wavelengths (i.e., colors) a sensor can in effect determine the temperature of the target it is tracking. Two-color tracking allows it to discriminate against the hotter flare and continue tracking the target. This creates a significant increase in the complexity required of countermeasures. To fool a two-color tracker, it is necessary to either use a large expendable object at the correct temperature or to fool the missile sensor in some other way to cause it to receive the proper energy ratio at the two measured wavelengths.

Flares have the disadvantage that they are expendable, therefore limited in number. Also, since they are very hot, they represent a significant safety hazard which prevents their use on civilian aircraft.

4.7.2 IR Jammers

IR jammers generate IR signals which attack the guidance signals passed to the sensors in IR-guided weapons. They provide IR signals similar to those produced by the IR energy of a target passing through a reticule as described in Section 4.2. When both the jamming signal and the modulated target energy signal are received by the missile's IR sensor, they cause the tracker to produce incorrect guidance commands.

Optimum use of an IR jammer requires information about the spin-and-chop frequencies of the missile seeker that is being jammed. This can be measured by scanning the missile tracker with a laser. The IR detector surface is reflective, and the lens gives the laser a double advantage (amplifying both on the way in and on the way back out). As the reticule moves over the sensor (see Section 4.2), the level of reflected signal will change. This allows a processor to reconstruct the waveform and phase of the energy pattern reaching the weapon's IR sensor.

Once the missile tracking signal is determined, the IR jammer can create an erroneous pulse pattern as in Figure 4.36 that will cause the missile tracking signals to produce incorrect steering commands in a manner similar to the effect of a deceptive RF jammer. Without direct information about the tracking in the specific attacking missile, generic false tracking signals can also be generated and transmitted.

The IR jamming signal comprises pulses of IR energy which can be generated in several ways. One way is to flash a Xenon lamp or an arc lamp. Another way, as shown in Figure 4.37, is the time-controlled exposure of a mass of hot material (called a "hot brick"). Mechanical shutters expose the hot brick to generate the required jamming signal. Both of these techniques produce jamming signals over a wide angular area for wide aspect protection.

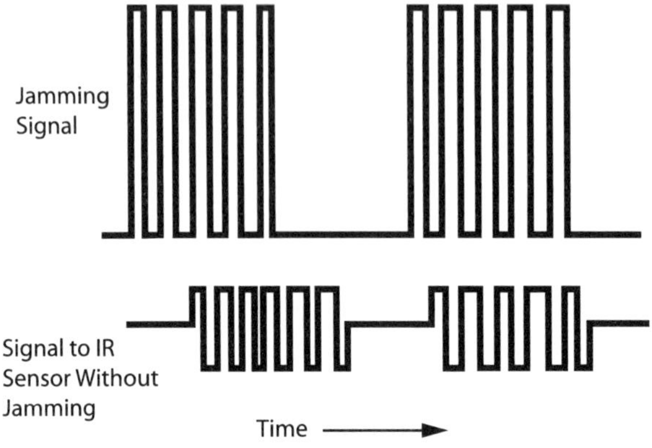

Figure 4.36 A stronger jamming signal combines with the IR signature of the target in the missile's IR receiver, distorting the angular tracking signals processed by the missile seeker.

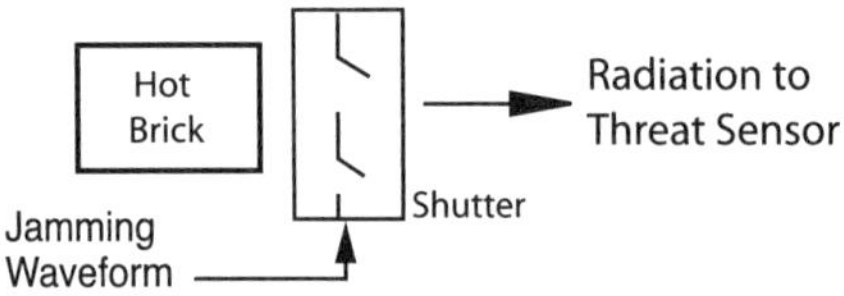

Figure 4.37 An IR jammer can produce its jamming signal by sequentially exposing a hot mass "hot brick" to the threat sensor by opening mechanical shutters.

A third way to generate the jamming signal is by use of an IR laser. The laser is easy to modulate and can produce very high level jamming signals, but is intrinsically narrow in beamwidth. Therefore, it must be accurately pointed at the tracker it is jamming. This requires beam steering controlled by an IR sensor with high angular resolution. The sensor typically detects the IR signature of the platform carrying the tracker (e.g., the missile). Because of its high signal levels at the receiver, the IR laser jammer can protect large platforms.

Note that if a jammer located on the protected target fails to deceive the missile tracker, it may act as a beacon to improve the missile tracking accuracy.

4.7.3 IR Decoys

IR decoys can be used to draw IR missiles away from protected platforms of any kind. Decoys can be fixed or can maneuver in ways to optimize deception of weapon trackers. Under some circumstances, they can be larger than flares to provide more energy at lower temperatures.

Decoys radiating the same order of magnitude of IR energy as a fixed or mobile ground asset can saturate enemy targeting capabilities.

4.7.4 IR Chaff

If material with a significant IR signature is deployed from an aircraft or from a rocket launched by a ship, it will have much the same protective capability against IR-controlled weapons as radar chaff provides against radar-controlled weapons. IR chaff can burn or smolder to create the proper IR signature or can oxidize rapidly to raise its temperature to the proper level. Since a chaff cloud occupies a large geometric area, it may be more effective against some kinds of tracking. Like RF chaff, the IR chaff can be used either to break a missile lock or to increase the background temperature to make target acquisition more difficult.

5

EW Against Communications Signals

In this chapter, we will deal with EW related to communications signals. Subjects will include radio propagation, the nature of the threat environment, and individual signal characteristics. It will also include discussions of search, intercept, and jamming issues relative to communication signals. The tactical communication environment is extremely dense in battlefield situations, which is an important consideration in all communications EW activities.

5.1 Frequency Ranges

Tactical communication is primarily conducted in the HF, VHF, and UHF frequency ranges as shown in Figure 5.1. However, there are also fixed point-to-point, satellite, and air-to-ground data links that must be considered communication signals as well. Table 5.1 lists the typical uses of communication links in each category.

In general, the higher the frequency, the more dependent a communication link is on a clear line of sight between the transmitter and receiver—but the more bandwidth is available. There are also unique considerations for each band.

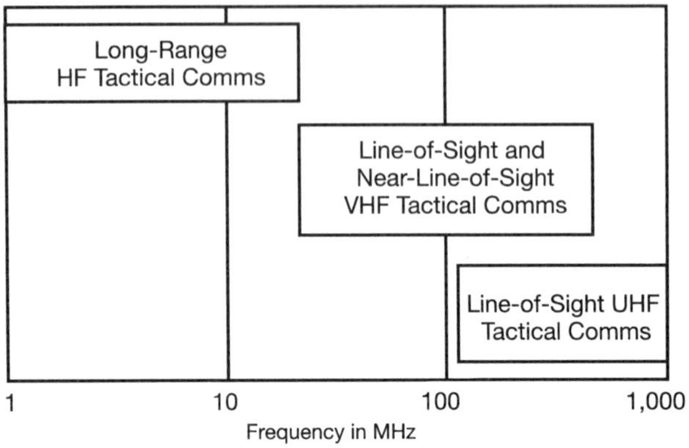

Figure 5.1 Tactical communications are typically conducted in the HF, VHF, and UHF frequency ranges.

Table 5.1
Communication Links

Military Application	Type of Link	Frequency Range
Tactical command and control (ground)	Ground point-to-point and air-to-ground	HF, VHF, UHF
Tactical command and control (air)	Air-to-ground and air-to-air	VHF, UHF
UAV command and data	Air-to-ground and air relay; satellite	Microwave
Strategic command and control	Satellite	Microwave

5.2 HF Propagation

This section is intended only to give general understanding of HF propagation, which is very complex. Its characteristics vary with time of day, time of year, location, and conditions (such as sunspot activity) that impact the ionosphere. An excellent article by Richard Groller in the June 1990 issue of the *Journal of Electronic Defense* ("Single Station Location HF Direction Finding") is suggested as a starting point for further study. The next suggestion is a handbook such as *Reference Data for Radio Engineers* (*RDRE*), which includes typical curves for HF propagation. Finally, for specific ionospheric conditions, propagation parameters, and other examples, the Federal Communication Commission has a Web site with loads of data (http://www.fcc.gov).

In this section, we will discuss the ionosphere, ionospheric reflection, HF propagation paths, and single-site locator operation. The primary references for this section are Mr. Groller's article and the *RDRE.*

HF propagation can be line of sight, ground wave, or sky wave. Where line of sight exists, propagation is predicted by the formulas presented for VHF and UHF propagation in Section 5.3. Ground wave, which follows the Earth, is a strong function of the quality of the surface along the path. The FCC Web site has some curves for this propagation mode. Beyond about 160 km, HF propagation depends on sky waves reflected from the ionosphere.

5.2.1 The Ionosphere

The ionosphere is a region of ionized gasses from about 50 to 500 km above the Earth's surface. Its primary interest here is that it reflects radio transmissions in the medium- and high-frequency ranges. As shown in Figure 5.2, the ionosphere is divided into several layers:

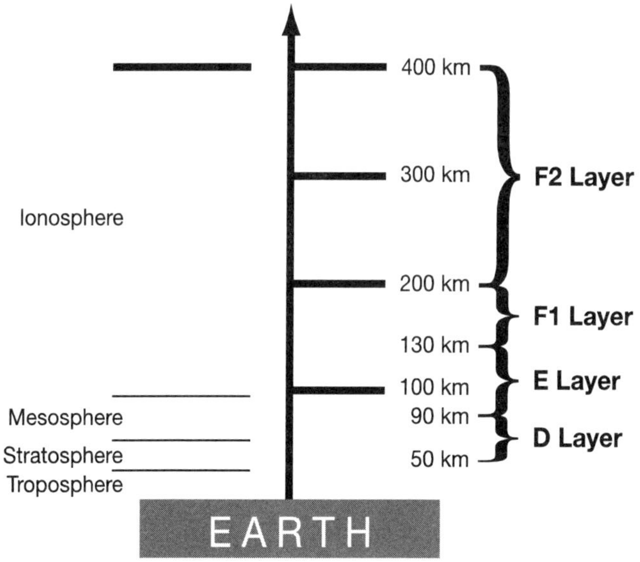

Figure 5.2 The ionosphere is characterized with D, E, F1, and F2 layers.

- The D layer is from about 50 to 90 km above the Earth. It is an absorptive layer, with absorption decreasing with frequency. Its absorption peaks at noon and is minimal after sunset.
- The E layer is from about 90 to 130 km above the Earth. It reflects radio signals for short- and medium-range HF propagation during the daytime. Its intensity is a function of solar radiation and varies with the seasons and sunspot activity.
- Sporadic-E is a condition causing a short-term transient layer of ionization that is present in local summer, primarily in Southeast Asia and the South China Sea. It causes short-term changes in HF propagation.
- The F1 layer extends from about 175 to 250 km above the Earth. It exists only during the daytime and is strongest during summer and periods of high sunspot activity. It is most prominent at the middle latitudes.
- The F2 layer extends from about 250 to 400 km above the Earth. It is permanent, but extremely variable. It allows long-range and nighttime HF propagation.

5.2.2 Ionospheric Reflection

Reflection from the ionosphere is characterized by virtual height and critical frequency. The virtual height as shown in Figure 5.3 is the apparent point of reflection of a signal from an ionspheric layer. This is the height measured by sounders which transmit vertically and measure the round-trip propagation

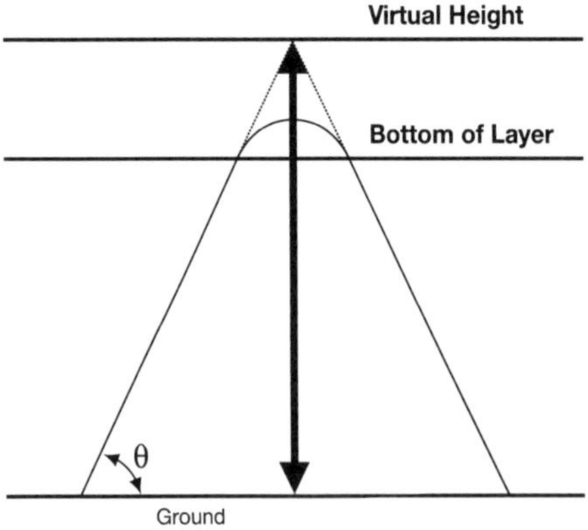

Figure 5.3 The virtual height of the ionosphere is the apparent reflection point for HF transmissions.

time. As frequency is increased, the virtual height increases until the critical frequency is reached. At this frequency, the transmission passes through the ionospheric layer. If there is a higher layer, the virtual height increases to the higher layer.

The maximum frequency at which reflection can occur is also a function of the elevation angle (θ in Figure 5.3) and the critical frequency (F_{CR}). The maximum usable frequency (MUF) is determined by the formula:

$$MUF = F_{CR} + \sec(\theta)$$

5.2.3 HF Propagation Paths

As shown in Figure 5.4, there can be several different transmission paths between a transmitter and a receiver depending on ionospheric conditions. If the sky wave passes through one layer, it may be reflected from a higher layer. There can be one or more hops from the E layer, depending on the transmission distance. If the E layer is penetrated, one or more hops from the F layer can occur. At night, this would be from the F2 layer and in the daytime from the F1 layer. Depending on the local density of various layers, there can also be hops from the F layer to the E layer, back to the F layer, and finally to the Earth.

The received power from sky wave propagation is predicted by the following formula:

$$P_R = P_T + G_T + G_R - \left(L_B + L_i + L_G + Y_P + L_F\right)$$

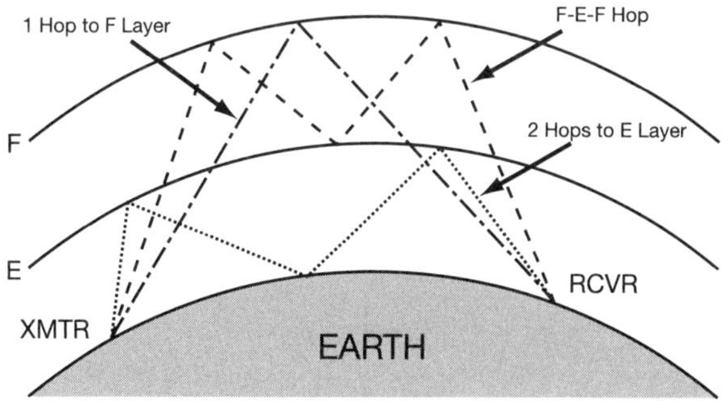

Figure 5.4 There are several possible propagation paths from a transmitter to a distant receiver, depending on the signal frequency, the ionospheric conditions, and the transmitter/receiver location geometry.

where P_T is the transmitter power, G_T is the transmit antenna gain, G_R is the receiving antenna gain, L_B is the spreading loss, L_i is the ionospheric absorption loss, L_G is the ground reflection loss (for multiple hops), Y_P is the miscellaneous loss (focusing, multipath, polarization), and L_F is the fading loss. All terms of the equation are in decibel form.

5.2.4 Single-Site Locators

A single-site locator (SSL) determines the location of an HF emitter by measuring the azimuth and elevation of arriving signals. The measured elevation is the angle of reflection from the ionosphere. As shown in Figure 5.5, the elevation angles at the transmitter and receiver are the same, and the distance from the SSL site to the emitter is given by the following formula:

$$D = 2R\left(\pi / 2 - B_R - \sin^{-1}\left(R \cos B_R / \{R + H\}\right)\right)$$

where D is the Earth surface distance from the SSL to the emitter, R is the radius of the Earth, B_R is the elevation angle measured at the receiver, and H is the virtual height of the ionospheric layer from which the signal is reflected.

5.2.5 Emitter Location from Airborne Systems

There have been a number of airborne electronic warfare and reconnaissance systems designed to intercept and locate HF transmitters using line-of-sight reception. Both the line-of-sight and sky wave signals reach the aircraft, but the

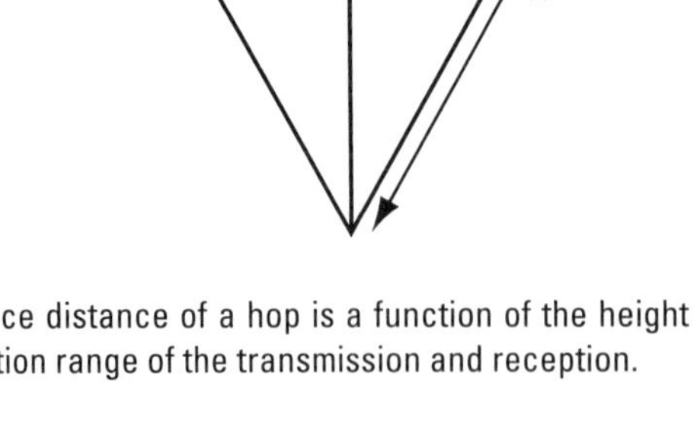

Figure 5.5 The surface distance of a hop is a function of the height of the ionosphere and the elevation range of the transmission and reception.

difference in path length causes severe multipath interference. This makes intercept difficult and causes significant problems with the operation of emitter location systems.

One solution to this problem is to use antennas which have gain pattern nulls at the top, for example, horizontal loops. Since line-of-sight emitters are relatively close to the aircraft subvehicle point, the sky wave signals come in at very high elevation angles. Thus, they are greatly attenuated by the antenna gain pattern.

5.3 VHF/UHF Propagation

Radio propagation in the VHF and UHF frequency ranges is "better behaved" than HF propagation. That is, the VHF and UHF propagation can be better described by formulas. In this section, we will discuss the commonly used propagation models and their normal application. Note that we are considering only the losses related to link geometry. There are additional losses from atmosphere and rain, but in this frequency range, they are usually not too significant.

5.3.1 Propagation Models

Chapter 84 (page 1182) of the *Communications Handbook* is an excellent reference for propagation models. It discusses both the simple and the complex models. It includes coverage of the Okumura, Hata, and Walfish and Bertoni models for outdoors propagation and the Saleh and SIRCIM models for indoor propagation. These models input specific path characteristics and thus provide valuable information for fixed-location communication, but they are less useful in electronic warfare. EW propagation typically deals with dynamic scenarios involving large numbers of actual and potential links. Therefore, we usually use either the free space, two-ray, or knife-edge diffraction propagation models in EW applications.

5.3.2 Free Space Propagation

As shown in Figure 5.6, the free space propagation (also sometimes called line-of-sight) model is appropriate for propagation where there are no significant reflection paths. This occurs at high frequencies and high altitudes, but also occurs when narrow-beam antennas reduce the impact of reflection paths on the propagation.

The propagation loss for free space propagation is given by the formula:

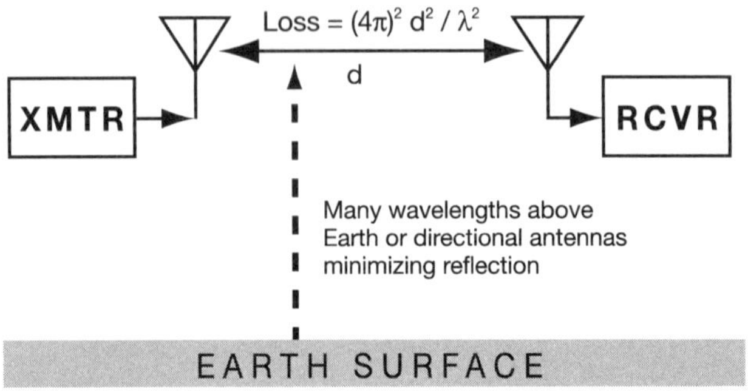

Figure 5.6 The free-space propagation model generally applies at high frequencies and/or high altitudes.

$$L = (4\pi)^2 d^2 / \lambda^2$$

where L is the direct loss ratio, d is the distance in meters, and λ is the transmission wavelength in meters.

The widely used decibel form of this equation is:

$$L = 32.44 + 20\log(f) + 20\log(d)$$

where L is the loss in decibels, f is the transmission frequency in megahertz, and d is the distance in kilometers.

Note that the constant 32.44 (commonly rounded to 32) requires that distance be input in kilometers. If distance in statute miles is used, the constant is 36.57 (commonly rounded to 37), and if nautical miles are used, the constant is 37.79 (commonly rounded to 38).

5.3.3 Two-Ray Propagation

In areas where propagation is characterized by a single reflection from the ground, the two-ray model is commonly used. As shown in Figure 5.7, this occurs in signals at lower frequencies and with transmissions close to the Earth's surface.

The propagation loss for two-ray propagation is independent of frequency. It is given by the formula:

$$L = d^4 / h_t^2 h_r^2$$

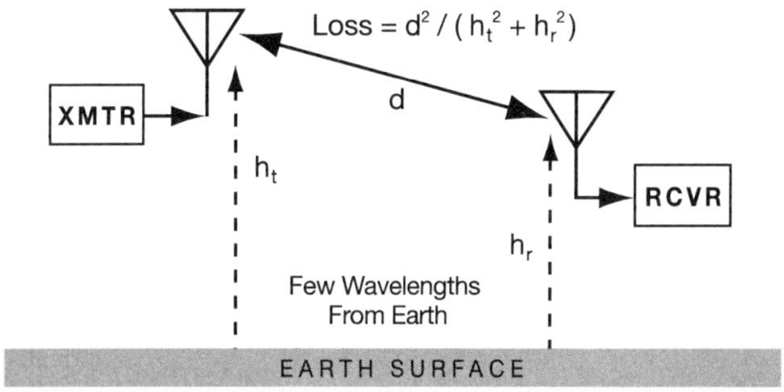

Figure 5.7 The two-ray propagation model generally applies at lower frequencies and altitudes.

where L is the direct loss ratio, d is the link distance in meters, h_t is the transmitting antenna height in meters, and h_r is the receiving antenna height in meters.

The more convenient decibel form of this equation is:

$$L = 120 + 40\log(d) - 20\log(h_t) - 20\log(h_r)$$

where L is the loss in decibels, d is the path distance in kilometers, h_t is the transmitting antenna height in meters, and h_r is the receiving antenna height in meters.

As shown in Figure 5.8, the determination as to which propagation model to use can be made by calculating the Fresnel zone (FZ). If the path distance is less than FZ, use the free space model, if the path is longer, use the two-ray model. The selected model is used for the whole path length. Note that when

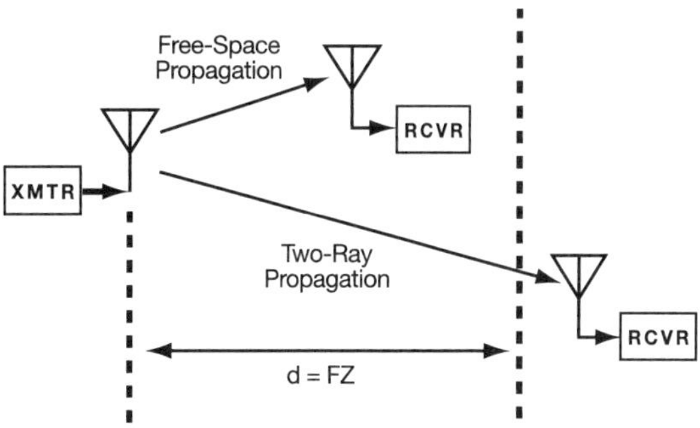

Figure 5.8 The Fresnel zone (FZ) distance determines whether the free-space or two-ray propagation model should be used over the whole transmission distance.

distance = FZ, the two models give the same propagation loss. The formula for calculating the Fresnel zone distance is:

$$FZ = 4\pi\, h_t h_r / \lambda$$

where FZ is the Fresnel zone distance in meters, h_t is the transmitting antenna height in meters, h_r is the receiving antenna height in meters, and λ is the wavelength of the transmitted signal in meters. Alternately, you can determine FZ from the formula:

$$FZ = (h_t\, h_r f) / 24{,}000$$

where FZ is in kilometers, the antenna heights are in meters, and f is the frequency in megahertz.

5.3.4 Knife-Edge Propagation

When the propagation path comes close to a ridge line, or does not quite clear it, there is a loss in addition to the propagation model losses described above. This loss is approximated by the knife-edge propagation model as shown in Figure 5.9. Note that the distance d_2 on the receiver side of the knife edge must be equal to or greater than the distance d_1 on the transmitter side. Also, that a distance term d must be calculated from the formula:

$$d = \left[\text{sqrt}(2) / \left(1 + d_1 / d_2\right)\right] d_1$$

In the example drawn in Figure 5.9, d_1 and d_2 are both 14 km, the straight-line transmission path passes 40m below the knife edge, and the transmission frequency is 150 MHz.

The d is calculated to be 10 km, so a line is drawn from 10 km through 40m to the index line. A second line is drawn from the index intersection of the first line through 150 MHz and on to the right-hand scale. This line passes through 10 dB, so there is 10-dB knife-edge loss in addition to the clear path propagation loss. Note that had the transmission path passed 40m above the knife edge rather than 40m below it, the knife-edge loss would have been 2 dB.

5.4 Signals in the Propagation Medium

When describing the communications link, we define the signal leaving the transmit antenna as effective radiated power in dBm. This is not literally true, since dBm is only defined inside a circuit. Out in the propagation medium

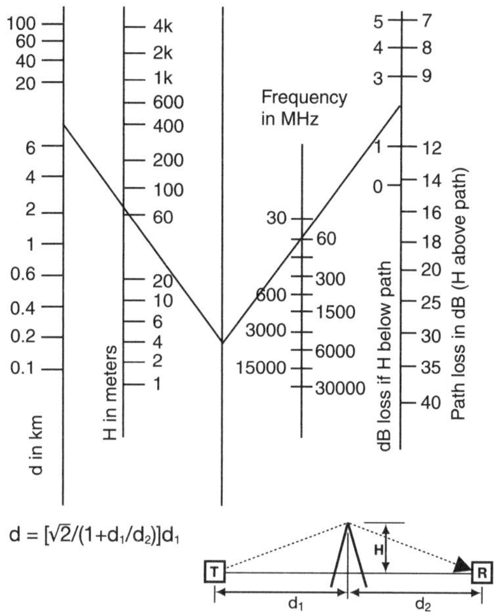

Figure 5.9 Knife-edge diffraction loss is a function of frequency and the location of the transmitter and receiver relative to the edge.

(atmosphere or space) the signal is accurately defined in terms of field strength, and the proper units are microvolts per meter. However, it is extremely convenient to describe signal levels in dBm through the whole link, so we use an artifice to make it work. The artifice is to describe the signal level in space as the power in dBm that would be received by an isotropic (omnidirectional) antenna as shown in Figure 5.10. If this ideal antenna receives the signal anywhere in the signal transmission path, the antenna output is properly in dBm.

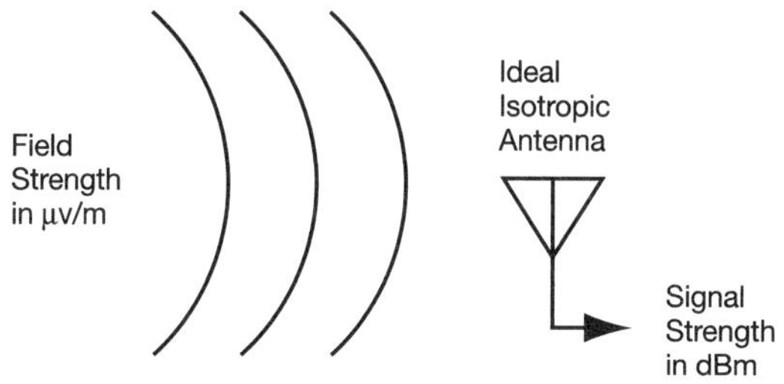

Figure 5.10 Field strength is converted to signal strength by calculating how it would be received by an ideal isotropic antenna.

Since receiver sensitivity and some other important items can be specified in microvolts per meter, we are sometimes required to convert back and forth between field strength and equivalent signal strength in dBm in order to work propagation problems. The conversion is done by squaring the field strength value, multiplying by the equivalent area of the isotropic antenna, and dividing by the impedance of free space.

The effective area of an antenna is given by the formula:

$$A = G\lambda^2 / 4\pi$$

where:

A = the antenna area in square meters;

G = the antenna gain (not in decibels);

λ = the signal wavelength in meters.

For an isotropic antenna, the gain is unity, so the effective area is simply $\lambda^2/4\pi$.

The decibel form equation for the antenna area is

$$A = 39 + G - 20\log(F)$$

where

A = the effective area in dBsm;

G = the gain in decibels;

F = the frequency in megahertz.

The 39 is a constant (in decibels) which includes the square of the speed of light, 4π and unit conversion factors. For an isotropic antenna the gain is unity (0 dB), so the effective area in dBsm is just 39 –20 log (F). The impedance of free space is 120π.

Multiplying the square of the field strength by the antenna area and dividing by the impedance of free space yields the formula:

$$P = E^2\lambda^2 / 480\pi^2$$

where

P = signal strength in watts;

E = field strength in microvolts per meter;

λ = wavelength in meters.

Note that the free space impedance (with units of ohms) is part of the denominator.

The decibel form of this equation is:

$$P = -77 + 20\log(E) - 20\log(F)$$

where

P = the power in the antenna output in dBm;

E = the field strength in microvolts per meter;

F = the frequency in megahertz.

The term –77 (dB) includes c^2, π^2, and unit conversion factors.

To convert from signal strength in dBm to field strength in microvolts per meter, use the formula:

$$E = \text{sqrt}(480\,\pi^2 P / \lambda^2)$$

where

E is the field strength in microvolts per meter;

P = signal strength in watts;

λ = wavelength in meters.

Note that ohms units are in the numerator.

The decibel form of this equation is:

$$E = \text{Antilog}\{[P + 77 + 20\log(F)]/20\}$$

where

E = field strength in microvolts per meter;

P = signal strength in dBm;

F = operating frequency in megahertz.

Table 5.2 shows the signal strength in dBm for various field strengths at various frequencies.

Table 5.2
Signal Strength in dBm for Various Values of Field Strength and Frequency

Field Strength (μv/m)	**Signal Strength @ 10 MHz**	**Signal Strength @ 50 MHz**	**Signal Strength @ 100 MHz**	**Signal Strength @ 250 MHz**	**Signal Strength @ 500 MHz**
1 μv/m	–97 dBm	–111 dBm	–117 dBm	–125 dBm	131 dBm
3 μv/m	–87.5 dBm	–101.4 dBm	–107.5 dBm	–115.4 dBm	–121.4 dBm
5 μv/m	–83 dBm	–97 dBm	–103 dBm	–111 dBm	–117 dBm
10 μv/m	–77 dBm	–91 dBm	–97 dBm	–105 dBm	–111 dBm
50 μv/m	–63 dBm	–77 dBm	–83 dBm	–91 dBm	–97 dBm
100 μv/m	–57 dBm	–71 dBm	–77 dBm	–85 dBm	–91 dBm

5.5 Background Noise

The chart in Figure 5.11 shows the background noise in various environments as a function of frequency. It is called external noise because it is not generated within the receiver. External noise is the combined emissions from many low-power interfering signals, such as engine spark plugs, trolley cars, and electric motors. Note that it is very strong in medium- and high-frequency (MF and HF) ranges and diminishes with increasing frequency.

- Atmospheric noise is mainly from lightning discharges. It is dependent on frequency, time of day, weather, season of the year, and geographical location.
- Cosmic noise is from the sun and the stars. It is highest in the Galactic plane.
- Urban and suburban noise is manmade noise from engine ignition, electric motors, electric switching, and high-tension line leakage.

The data in Figure 5.11 is from measurements made in a 10-kHz bandwidth with an omnidirectional antenna. If the received "noise" power is stated in microvolts per meter (as it is in some external noise graphs), it must be adjusted for the receiver bandwidth. Since this chart is in dBm above kTB (which includes a bandwidth term), it is valid for all bandwidths.

External noise enters the receiver through the receiving antenna as shown in Figure 5.12. (Note that kTB is generated inside the receiver.) If the receiving

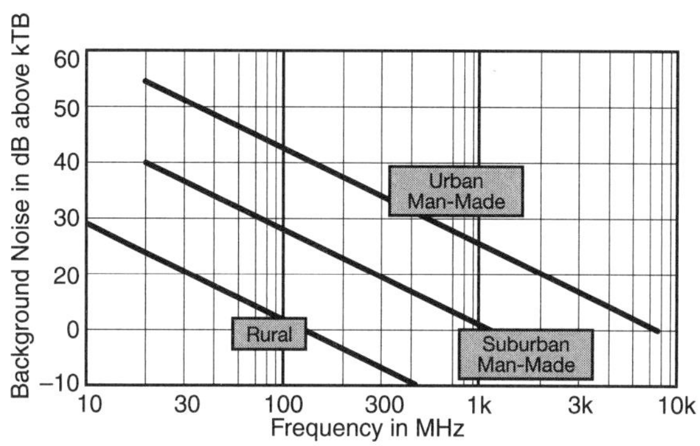

Figure 5.11 External-noise level is a function of frequency and the nature of the area in which the receiver is located.

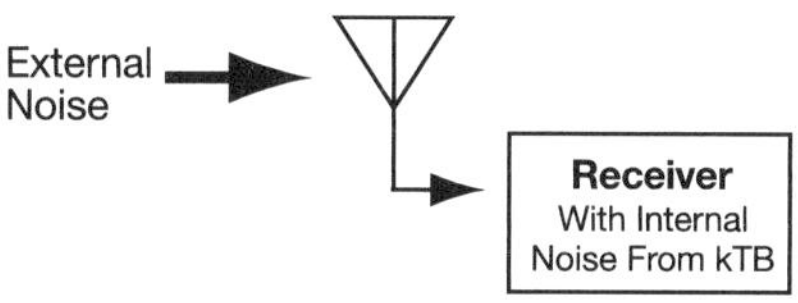

Figure 5.12 External noise adds to the internal kTB noise in a receiver to determine the signal-to-noise ratio achieved when a signal is received.

antenna is a whip, dipole, or similar antenna with 360° angular coverage, Figure 5.11 is applicable. With narrow-beam antennas, much lower levels are usually appropriate. The external noise is added to the internal kTB noise when determining the signal-to-noise ratio (SNR) achievable when a signal is received.

5.6 Digital Communication

More and more communication is becoming digital. Communication between computers is naturally digital, but current voice and video communication systems also typically transmit in digital form.

Digital communication is required for the use of high-security encryption and for some types of spread spectrum transmission. Digital signals can be efficiently multiplexed and routed to one or many intended recipients, and can be protected against intentional or unintentional interference by use of various error detection and correction approaches.

The good news about digital communication is that with reasonable care, the signal quality can be preserved through many changes of format, sequential link transmissions, and storage/retrieval cycles. The bad news is that once the signal is digitized, the output signal quality does not get any better through downstream processing—although processing can optimize the signal format and presentation characteristics for specific applications.

5.6.1 Digital Signals

A digital signal represents some information in digital form. Typically, this is in binary form—a series of "ones" and "zeros." Some transmitted information is inherently digital, for example when two computers communicate. When you hit a key on your PC keyboard, the computer generates an 8-bit signal to capture your keystroke. However, analog data is also converted to digital form for transmission. Typical digitized analog data includes voice signals, video signals (television, infrared scanner output or radar output) and instrumentation signals (temperature, voltage, angular position).

5.6.2 Digitization

Digitization is a complex field—which we cover here at a very rudimentary level to support later propagation and EW discussions. There are many excellent texts which cover digitization in much more detail, for example, Phillip Pace's book *Advanced Techniques for Digital Receivers.*

A "digitizer" is typically called an analog-to-digital converter, or ADC. As shown in Figure 5.13, digitization of an analog signal starts with the sampling of the signal. The rate at which the samples are taken determines the highest input signal frequency (or bandwidth) in the sampled signal that can be preserved in the digitized data. Then, the sampled value (i.e., signal amplitude) is digitized by comparing it to quantizing thresholds and generating a digital word which represents the highest quantizing threshold exceeded. Then, the sampling circuit is cleared to accept another sample. Finally, the digital signal is formatted for output to other circuits. This can either be in parallel or serial form. For digital transmission, it must typically be serial, with a series of sequential ones and zeros representing each sample.

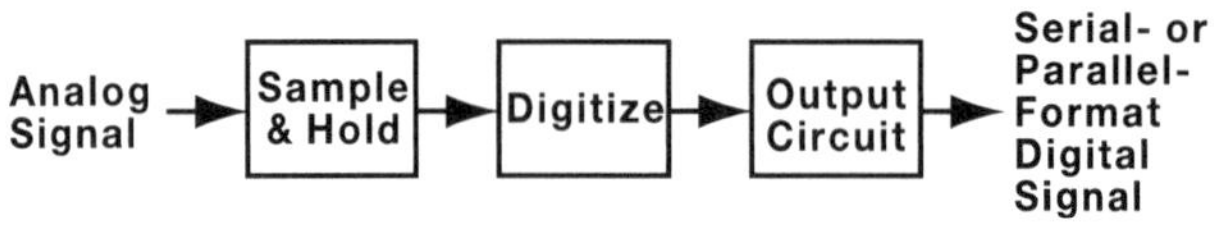

Figure 5.13 Digitization of an analog signal involves sampling, digitizing the amplitude of the sample, and formatting a serial or parallel digital signal.

Figure 5.14 shows an analog signal that is sampled at the points shown and is digitized to 4 bits of resolution or 16 quantization values. That is, the amplitude of the analog signal at each sample point is represented by a 4-bit digital word. Note that the analog curve is represented by the digital values: 0000, 0000, 0011, 0100, 0101. This figure makes several important points:

- The sampling rate determines the maximum frequency components of the curve captured. Another way of describing this effect is to say that the sampling rate must be fast enough to capture any character of the curve that you care about. Once the curve is digitized, it can only be accurately reconstructed to the stair step "reconstruction" curve. While you can filter this curve to knock off the corners, you can never make it more accurately represent the original analog data.
- The number of bits in each word limits the resolution with which the input amplitude is captured. This determines the "signal-to-noise ratio" and dynamic range of the reconstructed signal. Signal-to-noise ratio is in quotation marks because, although it is used in this context, it is really the signal-to-quantization ratio.
- The bits per sample also determine the dynamic range of the digitized signal. The dynamic range is the ratio between smallest signal that can be recovered in the presence of the largest signal present. Thus, if the system is configured to make the strongest signal full scale (i.e., 1111),

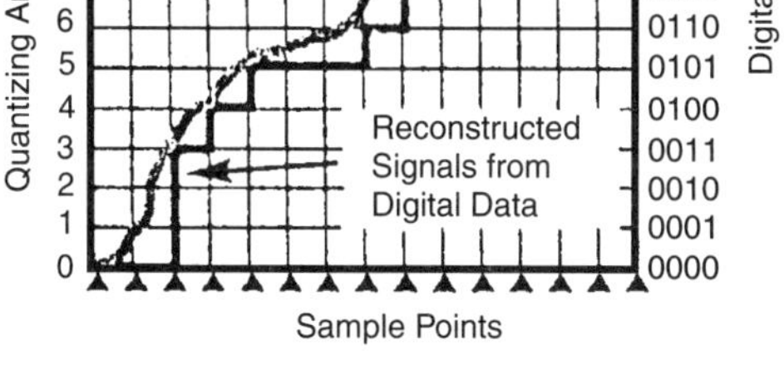

Figure 5.14 Once a signal is digitized, its accuracy is limited by the resolution to which it has been quantized.

the smallest signal captured must be at least 0001, or one "least significant bit" in amplitude.

Here are some formulas for these described values.

Maximum captured frequency or bandwidth versus sampling rate (called the Nyquist rate):

$$F_{max} \text{ or } BW_{max} = \frac{1}{2}(\text{sample rate})$$

Equivalent output signal-to-noise ratio:

$$\text{SNR} = 3 \times 2^{2m-1}$$

where SNR is not in decibels and m is the number of bits per sample. In decibel form, the equation is:

$$\text{SNR(dB)} = 5 + 3\,(2m - 1)$$

Dynamic range

$$DR = \left(2^m\right)^2$$

where DR is the dynamic range not in decibels and m is the bits per sample. In decibel form, the equation is:

$$DR(\text{dB}) = 20 \log_{10}\left(2^m\right) = 20 \log_{10}(\text{number of quantizing levels})$$

5.6.3 Digitizing Imagery

As shown in Figure 5.15, an image area is typically scanned by a TV camera or similar device to look at the whole screen, one tiny spot at a time. Each spot that

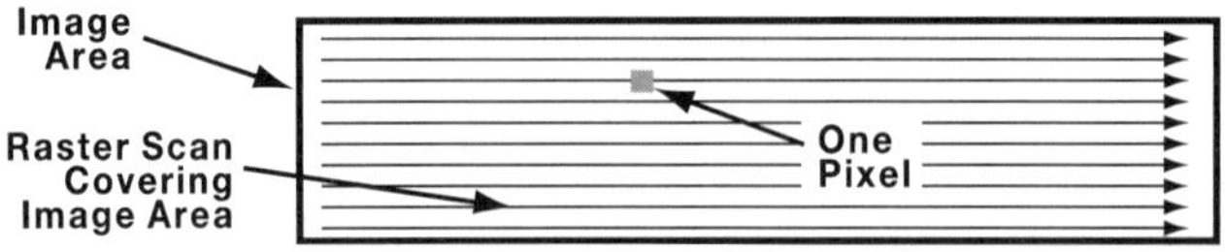

Figure 5.15 Any image can be digitized sequentially by covering it with a raster scan and digitizing the values observed in each pixel.

is captured is a pixel. The value of the signal at each pixel is digitized; for example, a TV camera captures the red, green, and blue amplitude at each pixel. Thus the digitized signal has a digital word for each color amplitude at each pixel.

5.6.4 Digital Signal Format

When digital data is transmitted, it requires additional information to allow its proper recovery.

- Since the transmitted data is a continuous stream of ones and zeros, there must be a synchronization scheme to allow the receiving equipment to determine which bit is which. After synchronization, the transmission is organized into frames and subframes. Synchronization need only be performed at the beginning of a transmission or periodically after several frames.
- Address bits direct data to the correct receiver when the data stream contains data for more than one purpose, for example, a message to a particular operator or data to a particular readout.
- Data bits contain the information that is to be transmitted (a segment of voice, imagery, computer data).
- Parity bits are added to detect incorrect bits that are introduced in the transmission process. There are also approaches through which incorrect bits can be corrected at the receiver. We will discuss these in detail later in this chapter.

Figure 5.16 shows a typical format for the transmission of digital data. The transmitted bits other than data bits are often called "overhead." Typical numbers for overhead are 10% to more than 50% of the full number of bits transmitted.

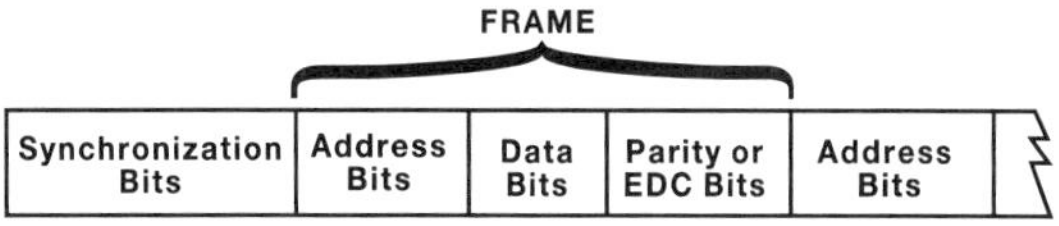

Figure 5.16 Transmitted digital data must be accompanied by overhead bits for synchronization and often by additional overhead bits for addressing and error reduction.

5.6.5 RF Modulations for Digital Signals

In order to be transmitted, digital data must be modulated onto an RF or laser carrier signal. We will focus on RF modulation, although the same discussion applies to the transmission at infrared or optical wavelengths. The purpose of this section is not to thoroughly cover digital modulations, but rather to provide enough information to support our later discussions of the detection, jamming, and interception of digital threat signals, and the protection of friendly digital signals.

5.6.5.1 Digital Modulations

Figure 5.17 shows an example of an simple digital modulation. This is an amplitude shift keyed (ASK) modulation. The RF carrier is amplitude modulated to carry the digital data. In this case, during transmission of a "one," the amplitude is higher than it is during a "zero." When this signal is AM detected in a receiver, the video output is evaluated against a threshold to regenerate the original digital signal. There are, of course, typically many more cycles of the RF carrier during each transmitted bit.

Similarly, frequency shift keying (FSK) transmits a "one" on one frequency and a "zero" on another frequency. FSK can be either coherent (with both frequencies derived from a single oscillator) or noncoherent.

Phase modulation is also widely used in digital communication. This modulation requires that the receiver have a reference oscillator. Then it can determine if the incoming signal is in phase with that oscillator, or if not, the instantaneous relative phase of the received signal. Figure 5.18 shows a BPSK signal. Since it is binary, there are two transmitted phases, 180° apart. Note that phase modulation must, by definition, be coherent.

There can also be phase shift keyed signals with more than two defined phases. In Figure 5.19, a QPSK signal is shown. It has four defined relative phases (0°, 90°, 180°, and 270°). Each of these phases represents two bits of data. In the figure, 0° represents "one, one," 90° represents "one, zero," 180°

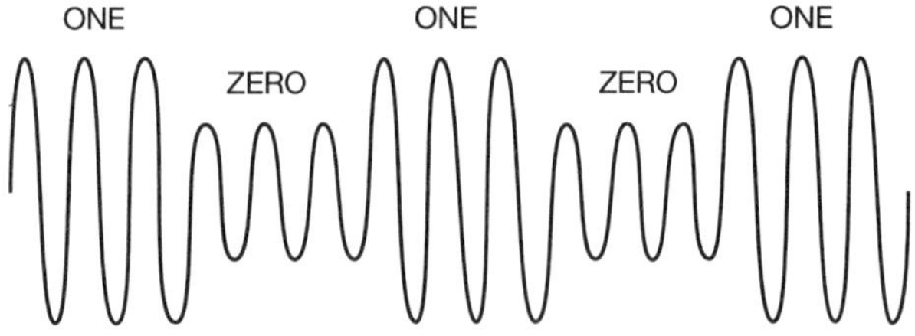

Figure 5.17 In an amplitude shift keyed (ASK) signal, the FR waveform is amplitude modulated to carry the digital modulation.

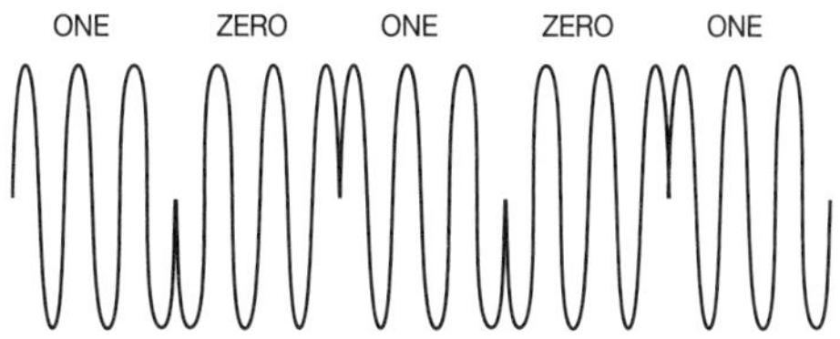

Figure 5.18 In a binary phase shift keyed (BPSK) signal, the FR waveform has one phase to carry 1 and a 180° different phase to carry 0.

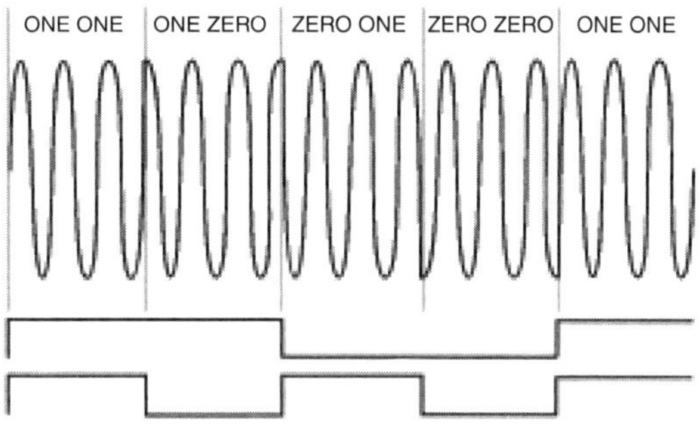

Figure 5.19 In a quadrature phase shift keyed (QPSK) signal, the FR waveform can have four phases. Each phase represents 2 bits of digital data.

represents "zero, one," and 270° represents "zero, zero." Each of the periods during which a signal phase is transmitted is called a "baud." In the earlier examples, one "bit" (of the digital data) was carried by one baud. In QPSK, there are 2 bits per baud. There are also more complex phase modulations in which there are many defined phases, so there can be more bits per baud carried. For example if there were 32 defined phases, there would be 5 bits transmitted with each baud. This is called "32-ary PSK."

5.6.5.2 Frequency Efficient Digital Modulations

In these examples, the signal waveform is shown moving instantaneously between the "one" and "zero" modulation states. When this is done, a great deal of RF bandwidth is required to carry the modulation (to handle the frequency content of the rapid transitions). There are also many frequency efficient modulations in which the transition between states is made more gently. Examples are sinusoidal phase shift keying, minimum shift keying, and others. We will discuss these in a little more detail in Section 5.6.8 when we are discussing the implication of the required bandwidth on signal transmission performance.

5.6.6 Signal-to-Noise Ratio

As discussed earlier, the signal-to-noise ratio of a reconstructed digitized signal is actually the signal-to-quantization-noise ratio. It is a function of the number of quantizing bits characterizing the signal waveform. In all our earlier discussions of signal transmission, we have defined the sensitivity of the receiver in terms of the required predetection signal-to-noise ratio (which we have called the RFSNR). (Remember that in much of the technical literature, this is also called the CNR.) For digital signals, the output SNR from the receiver is set by the original digitization of the information. However, we cannot recover that digital data unless the bits received are the same as the bits transmitted.

Incorrectly received bits are called bit errors, and the ratio of incorrect bits to total transmitted bits is called the "bit-error rate," which varies with E_b/N_0. E_b is the energy per bit, and N_0 is the noise per hertz of bandwidth. This function is the RFSNR modified by the bit rate and the effective receiver bandwidth. The formula is:

$$E_b/N_0 = \text{RFSNR} \times \text{bandwidth/bit rate}$$

or in decibel form:

$$E_b/N_0\,(\text{dB}) = \text{RFSNR}\,(\text{dB}) + 10\log(\text{bandwidth / bit rate})$$

In any digital transmission system, a required bit-error rate is specified. However, it may be stated in different forms. For example, it might be "one incorrect standard message per hour." This is converted to bit-error rate by considering the number of bits in a standard message and the rate at which messages are sent.

5.6.7 Bit-Error Rate Versus RFSNR

Received signal noise causes variations in the received modulation levels–for example causes the phases of the QPSK signal to be other than the exact 0°, 90°, 180°, or 270° values. This means that when the receiver is determining what phase is present, it will occasionally make a mistake. The lower the SNR (noise being random), the higher the ratio of errors.

For each type of modulation used to carry digital signals, there is a curve like that shown in Figure 5.20. It shows bit-error rate versus E_b/N_0 in the receiver; 10^0 is 1, so that would be 100% errors, and 10^{-6} is one error per million received bits. This figure actually shows the curve for noncoherent frequency

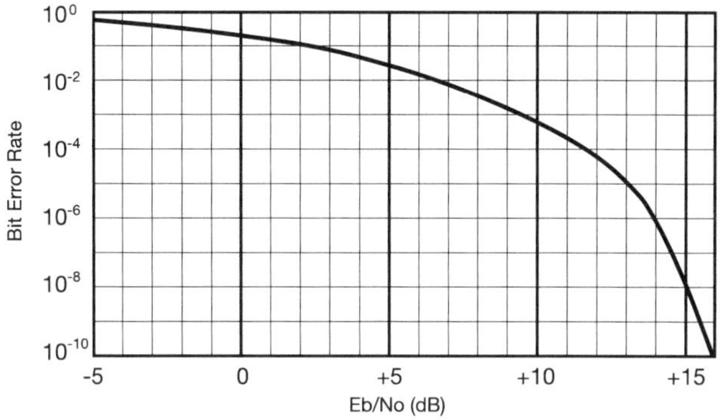

Figure 5.20 The predetection signal-to-noise ratio of the digitally modulated RF signal determines the bit-error rate in the digital data recovered from the signal. A curve like this is defined for each type of RF modulation used to carry digital signals.

shift keying, but is typical. All of these curves have about the same shape, but they would intersect the bottom of the curve at different E_b/N_0 values. A characteristic of all of these curves, is that they asymptotically approach 50% errors as the E_b/N_0 becomes very large.

The SNR values at which the curves for various modulations would intersect the bottom of our curve cover a range of more than 20 dB. An important generality is that the coherent form of a particular modulation (i.e., coherent FSK versus noncoherent FSK will provide the same bit-error rate at about 1 dB less SNR. Note that the difference between the required E_b/N_0 to achieve a particular bit-error rate with phase shift keying (antipodal signaling) is 3 dB less than required for frequency shift keying (orthogonal signaling).

These curves are also shown as a function of predetection signal-to-noise ratio (RFSNR). When the curve is drawn this way, there is an assumption of the ideal ratio of bit rate to bandwidth.

5.6.8 Bandwidth Required for Digital Signals

The frequency spread of digitally modulated radio frequency signals has a characteristic shape which depends on the bit rate of the transmitted data. The higher the data rate, the greater the frequency bandwidth required for its transmission.

5.6.8.1 The Digital Modulation Frequency Spectrum

Figure 5.21 shows the transmitted RF spectrum of a BPSK signal as it would be displayed on a spectrum analyzer. The main lobe of the frequency response is between the two nulls at the edges of the picture.

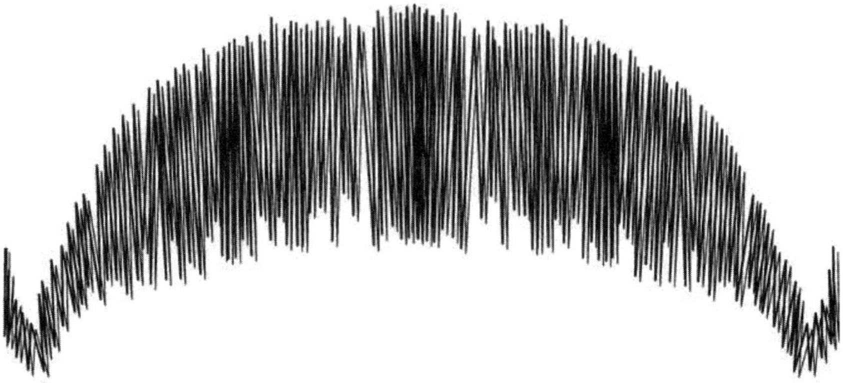

Figure 5.21 The frequency spectrum of a digitally modulated RF signal has a characteristic shape dictated by the bit rate.

This signal has a sin X / X frequency pattern. As shown in Figure 5.22, the main lobe is "twice the bit rate" wide (converted to radio frequency at 1 Hz per bps). The frequency side lobes are half as wide as the main lobe, and roll off with frequency distance from the main lobe. Only the main lobe and first side lobes are shown in this figure. The bandwidth of a digital signal is often taken to be between the frequencies at which the main lobe energy is 3 dB below the peak value (at the carrier frequency). However, the shape of the received bits depends on the transmission of higher frequency components, so more transmission bandwidth may be required.

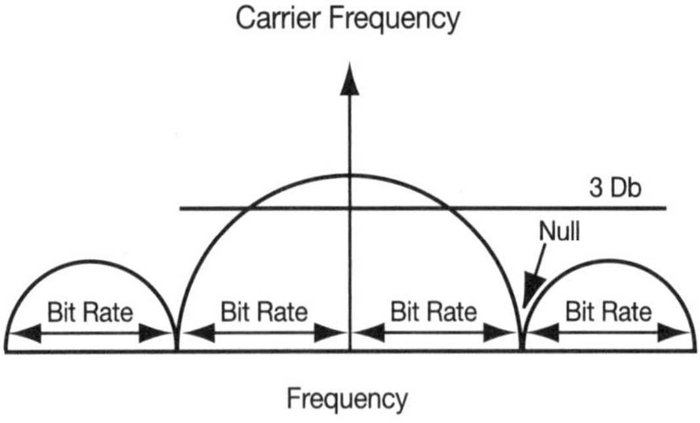

Figure 5.22 The null-to-null frequency bandwidth of a typical digitally modulated signal is twice the bit rate, converted at 1 Hz/bps.

5.6.8.2 Spectrum Expanding Signal Features

It is normally necessary to add synchronizing, address and parity bits to the digital bit stream in addition to the basic data carried. These extra bits are called the overhead and are often 10% to 20% of the transmitted signal. The bandwidth of the transmitted signal is determined by the actual transmitted bit rate, including these extra bits.

It is sometimes appropriate to encode digital data in error detection and correction (EDC) codes which allow bit errors introduced during transmission to be corrected in the receiver. A typical such code is the 31/15 Reed Solomon code used in the widely employed "Link 16" system (also called JTDS). This code transmits 31 bytes for each 15 bytes of data passed. Thus, the transmitted bit rate is more than doubled—proportionally increasing the required transmission bandwidth.

5.6.8.3 Frequency Efficient Modulations

In Section 5.6.5, we discussed ASK, BPSK, and QPSK modulations. These were described only in terms of the way that the modulated signal carries "one" and "zero" values. The modulation is presumed to move very quickly between the two values, which causes significant higher frequency components to appear in the modulation. This in turn causes the side lobes of the frequency spectrum to carry significant energy.

Figure 5.23 shows two modulations designed to reduce the higher frequency components, allowing a higher quality signal to be carried in less bandwidth. The two waveforms shown in the time domain are sinusoidal shift keying and minimum shift keying. The sinusoidal shift keying moves the signal between binary values sinusoidally, and the minimum shift keying moves it along a spectrum minimizing curve.

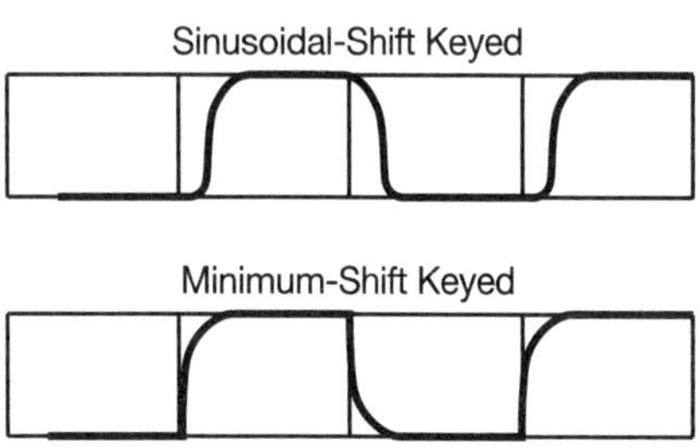

Figure 5.23 There are frequency-efficient digital modulations that move between 1 and 0 modulation values in a way that reduces the level of higher-frequency components.

Table 5.3 compares the frequency spreading of the minimum shift keyed signal with that of other digital modulation waveforms. This data comes from an excellent textbook by Robert Dixon, *Spread Spectrum Systems with Commercial Applications.* Note that the 3-dB bandwidth of BPSK, ASK, and QPSK signals is 88% of the clock rate—compared with twice the clock rate for the null-to-null bandwidth of the main lobe. For MSK signals, you will note that the null-to-null and 3-dB bandwidths are only 75% of that for conventionally modulated signals carrying the same data rate. You will further note that the power in the sidelobes is significantly reduced.

5.6.9 Impact of Signal Bandwidth on Electronic Warfare

There are several ways that the bandwidth of transmitted digital signals impacts electronic warfare. The most obvious is that the receiver bandwidth is a major determinant of sensitivity. The sensitivity of a receiver is the minimum signal that the receiver can receive and still do its job. Sensitivity is the sum of kTB, the receiver noise figure, and the required SNR. In the atmosphere, kTB (the thermal noise in the receiver) is calculated by the formula:

$$\text{kTB} = -114\ \text{dBm} + 10\log\left(\text{bandwidth} / 1\ \text{MHz}\right)$$

Thus, wider bandwidth, by reducing the sensitivity, requires additional transmitter power to provide adequate communication over any given operating range.

When receiving a hostile signal (in an ES or ELINT system), the sensitivity determines the range from which adequate interception and emitter location can be accomplished.

Another, more subtle, effect has to do with low probability of intercept (LPI) features used to protect communication systems. LPI features involve the

Table 5.3
Comparison of the Frequency-Spread Digital-Modulation Waveforms

Waveform	Null-to-Null Main Lobe BW	3 dB Bandwidth	First Side Lobe	Roll-Off Rate
BPSK, ASK, QPSK	2 × Code Clock	0.88 × Code Clock	–13 dB	6 dB/Octave
MSK	1.5 × Code Clock	0.66 × Code Clock	–23 dB	12 dB/Octave

deliberate spreading of the transmitted frequency bandwidth. The greater the LPI spreading factor, the harder it is to intercept, locate, or jam the signal. Since high data rate digital signals occupy large bandwidths, it is difficult to achieve large LPI spreading ratios before running into limitations in the bandwidth of amplifiers and antennas. Note that there is a discussion of LPI signals in Section 5.7.

Error-detection codes reduce the impact of certain types of jamming on digital signals. Although an error-correction code can improve the SNR in a communication receiver, the extra bandwidth required for the code will, except in low bit-error-rate applications, hurt performance more than the EDC SNR improvement will help it. However, when there is some specific error-inducing element in the communication situation—typically either jamming or a severe impact from even low bit-error rates—EDC can be used to great advantage. We will be discussing these codes and their impact on communication jamming in more detail later in this chapter.

5.7 Spread Spectrum Signals

In this section, we will review LPI signals as a prelude to our discussions about jamming them. These signals spread their energy (pseudorandomly) across a wider frequency range than required to just carry their information from the transmitter to the receiver. Thus, they are also called spread spectrum signals. The minimum bandwidth required for a communication transmission is the "information bandwidth." The "transmission bandwidth" is the frequency over which the signal is spread.

The desired receiver for a spread spectrum signal has dispreading capability that is synchronized to the spreading circuitry in the transmitter—allowing the receiver to process the signal in its original, unspread form as shown in Figure 5.24. A hostile receiver does not have the synchronized dispreading

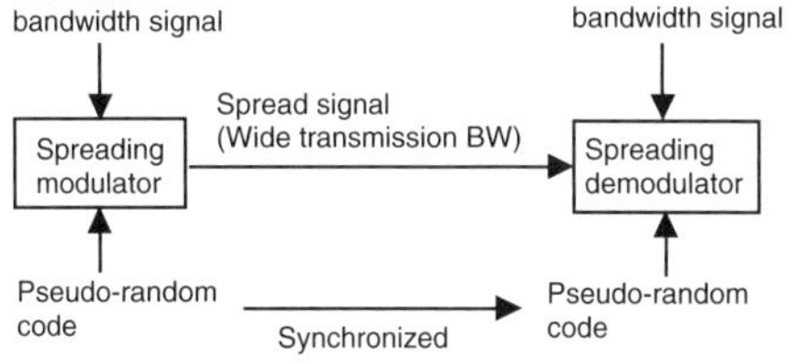

Figure 5.24 Frequency spreading schemes transmit at a bandwidth much wider than the information bandwidth of the signal being transmitted.

capability. Thus, signal intercept and jamming and emitter location are greatly complicated. The noise power in a receiver is proportional to its effective bandwidth. Thus, the noise power in a hostile receiver with enough bandwidth to receive the spread spectrum signal will be high enough to hide the signal.

There are three basic types of spread spectrum signals: frequency hopping, chirp, and direct sequence. Each approach spreads the signal. However, the nature of the power versus frequency versus time distribution for each type of modulation gives it different vulnerability to intercept, location, and jamming.

5.7.1 Frequency-Hopping Signals

As shown in Figure 5.25, frequency-hopping (FH) signals periodically move the information carrying signal to a different, randomly selected, transmission frequency. The desired receiver hops with the transmitter, but a hostile receiver does not know the hopping sequence. The hopping period is typically less than 10 ms, and can be much shorter. FH is an important technique for military communication since the signal spread can be very wide.

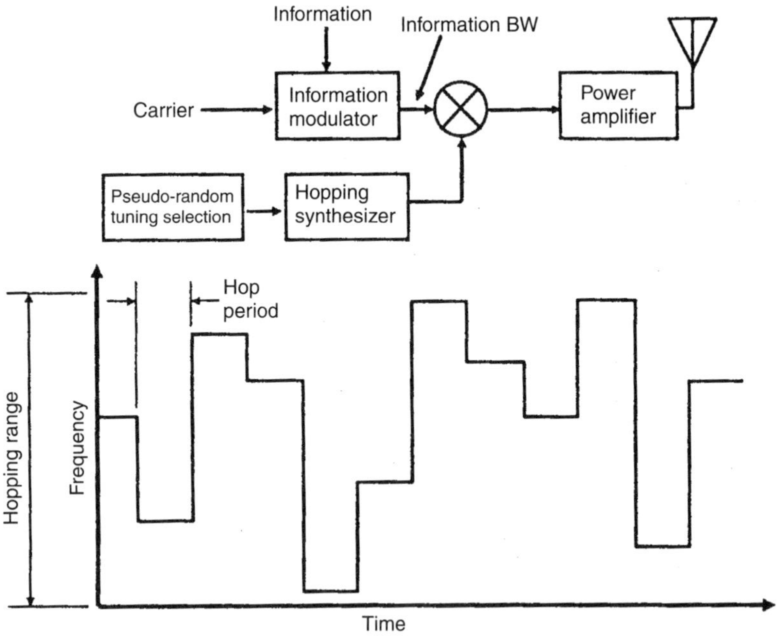

Figure 5.25 Frequency-hopped signals hop between randomly selected frequencies over a wide frequency range.

Frequency hoppers must carry digitally modulated information so that the output signal can be retimed to avoid dropouts while the hopping synthesizer (shown in Figure 5.25) settles on each new transmitting frequency. A "slow hopper" transmits multiple data bits per hop, and a "fast hopper" switches frequencies multiple times per data bit. Most current FH systems are slow hoppers; fast hoppers require very sophisticated synthesizers.

5.7.2 Chirp Signals

Figure 5.26 shows the spreading modulation of a chirp signal and the way the signal is generated. Detection of a signal normally requires that the signal remain within a receiver for at least the inverse of the effective receiver bandwidth (e.g., 1 μs if the bandwidth is 1 MHz). The chirp transmitter tunes much faster than that, so a receiver with sufficiently narrow bandwidth to detect the signal will not see it long enough to detect it, and if wide enough to accept the whole chirp range will have inadequate SNR.

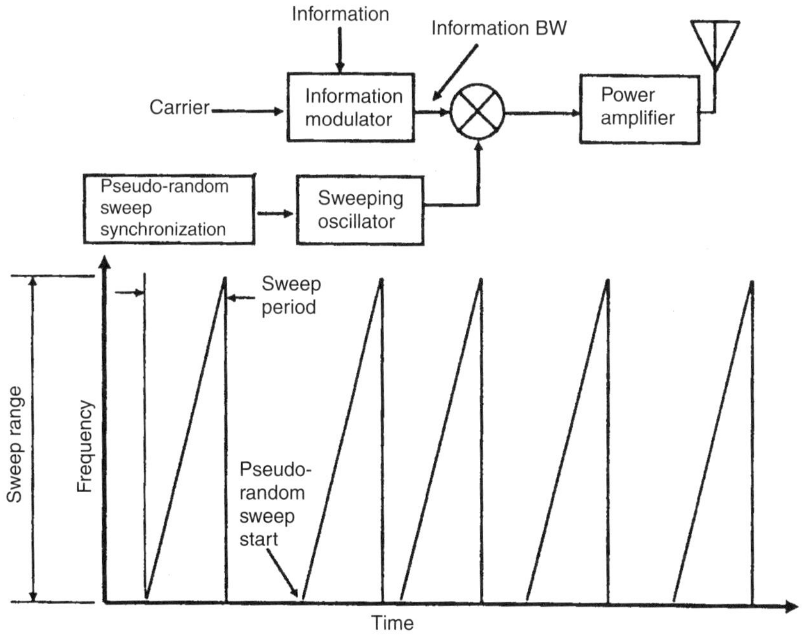

Figure 5.26 Chirp signals are rapidly swept over a frequency much wider than the information bandwidth of the transmitted signals.

The desired receiver sweeps in synchronization with the transmitter, and can therefore use bandwidth near the information bandwidth. The start time of each sweep can be random and/or the slope of the sweep can be nonlinear to make it difficult for a hostile receiver to synchronize with the transmitter.

5.7.3 Direct Sequence Spread Spectrum Signals

As described in Section 5.6.8, the frequency occupancy of a digital signal is proportional to the bit rate. If the digital signal is modulated a second time with a much higher bit rate, as in Figure 5.27, the signal energy is spread across a proportionally larger frequency range. This process is called direct sequence spread spectrum (DSSS) modulation. The bits of the spreading waveform are called "chips." As shown in the figure, the signal continuously occupies a much larger spectrum. Actually, this part of the figure is a little deceptive, since the actual spectral distribution will be the sinx/x curve shown in Figure 5.22 for any digital signal. The null-to-null bandwidth of the information carrying digital signal will

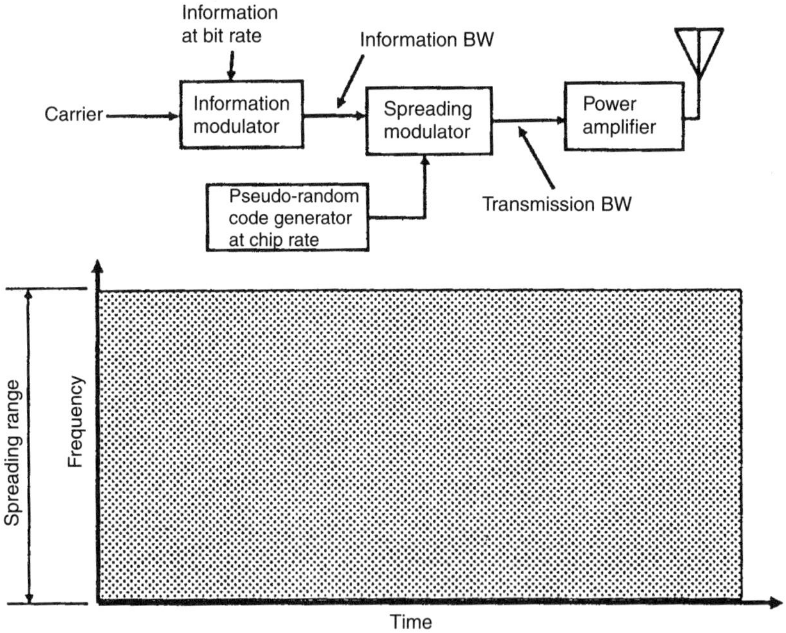

Figure 5.27 A direct sequence spread spectrum signal is continuously spread over a frequency range much wider than the information bandwidth of the transmitted signal.

be twice the bit rate, while the null-to-null bandwidth of the DSSS signal will be twice the (much higher) chip rate.

5.8 Communications Jamming

The primary difference between radar and communication jamming is in the geometry. Figure 5.28 shows the communication jamming geometry. Whereas a typical radar has both the transmitter and the associated receiver at the same location, a communication link, because its job is to take information from one location to another, always has its receiver in a different location from that of the transmitter.

Note that you can only jam the receiver. Of course, communication is often done using transceivers (each including both transmitter and receiver), but only the receiver at location B in the figure is jammed. If transceivers are in use and you want to jam the link in the other direction, the jamming power must reach location A.

There are some important communications cases in which transceivers are not used, for example, in UAV links as shown in Figure 5.29. This figure shows the data link (or "downlink") being jammed. Again, you jam the receiver.

Another difference from radar jamming is that the radar signal makes a round trip to the target, so the received signal power is below the transmitted power by the fourth power of the distance (often stated as –40 log range). Since the jammer power is transmitted one way, it is only reduced by the square of the distance. In communication jamming, both the desired transmitter

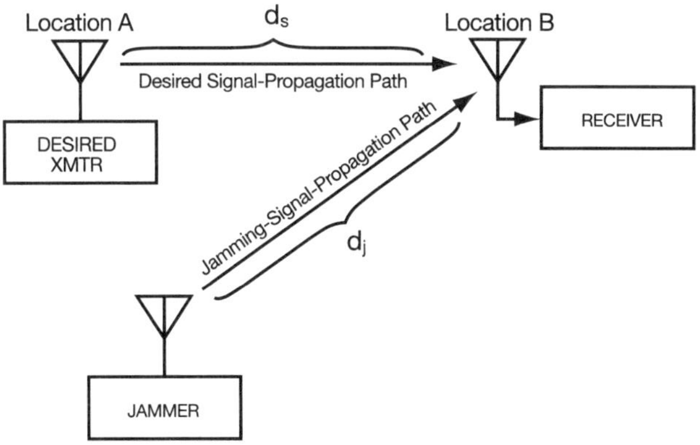

Figure 5.28 The communications-jamming geometry has one-way links from the desired transmitter and the jammer to the receiver.

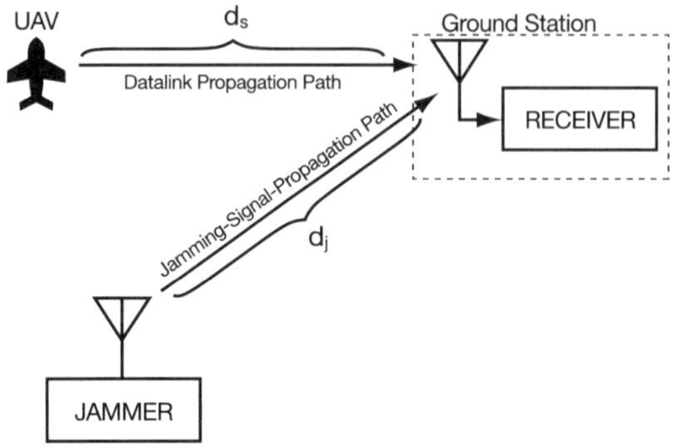

Figure 5.29 A jammer operating against a UAV data-link must jam the receiver at the ground station.

power and the jammer power are reduced by the square of their respective distances.

5.8.1 Jamming-to-Signal Ratio

The jamming-to-signal (J/S) ratio equation for communication jamming is:

$$\mathrm{J/S} = \left(ERP_J\right)\left(G_{RJ}\right)\left(d_S^2\right) / \left(ERP_S\right)\left(G_R\right)\left(d_J^2\right)$$

where

ERP_J = the effective radiated power of the jammer (in any units);

ERP_S = the effective radiated power of the desired transmitter (in the same units);

d_J = the distance from the jammer to the receiver (in any units);

d_S = the distance from the desired transmitter to the receiver (in the same units);

G_{RJ} = the gain of the receiving antenna toward the jammer (not in decibels);

G_R = the gain of the receiving antenna toward the desired transmitter (not in decibels).

or in decibel notation:

$$\mathrm{J/S} = ERP_J - ERP_S + 20\log\left(d_S\right) - 20\log\left(d_J\right) + G_{RJ} - G_R$$

The terms are the same as above, but the ERP terms are both in units of either dBm or dBW, and the gains are in decibels.

In both equations, ERP is the effective radiated power toward the receiver. This is the product (or sum in the decibel equation) of the transmitter output power and the transmit antenna gain in the direction of the receiver.

In tactical communication, with all parties using transceivers with whip antennas, the gain of the receiving antenna is symmetrical in azimuth. Thus, the gain toward the jammer will be the same as the gain toward the desired receiver, so the last two terms (G_{RJ} and G_R) cancel each other.

5.8.2 Operation Near the Earth

Both of the above equations are based on the line-of-sight propagation loss model, which assumes that both the transmitter and receiver are several wavelengths from the ground. Thus, the spreading loss for each signal is given (in decibel form) by:

$$L_S = 32.44 + 20\log(d) + 20\log(F)$$

This is the line-of-sight loss equation described in Section 5.3.2 (with d in kilometers and F in megahertz).

In Section 5.3.3, there are equations for the spreading loss where there is one significant reflector (i.e., water or the Earth). If the transmitter is less than the Fresnel zone distance from the receiver, the line-of-sight loss equation applies. This occurs at high frequencies, with many wavelengths of antenna height and/or when narrow-beam antennas prevent significant reflections from flat Earth.

If either the jammer or the desired transmitter is beyond the Fresnel zone, the two-ray propagation loss model applies:

$$L_S = d^4 / h_t^2 h_r^2$$

where L_S is the direct loss ratio, d is the link distance in meters, h_t is the transmitting antenna height in meters, and h_r is the receiving antenna height in meters.

The more convenient decibel form of this equation (same units for d and h) is:

$$L_S = 120 + 40\log(d) - 20\log(h_t) - 20\log(h_r)$$

where L is the loss in decibels, d is the link distance in kilometers, and the antenna heights are in meters.

The J/S ratio will be proportional to the appropriate propagation loss models. For example, if both the jammer and the desired transmitter are beyond the Fresnel zone, the J/S equation is:

$$J/S = (ERP_J)(d_S^4)(h_J^2)(G_{RJ}) / (ERP_S)(d_J^4)(h_S^2)(G_R)$$

and the decibel form of the equation is:

$$J/S = ERP_J - ERP_S + 40\log(d_S) - 40\log(d_J)$$
$$+20\log(h_J) - 20\log(h_S) + G_{RJ} - G_R$$

The units in both equations are the same as for the line-of-sight equations, with the addition of meters for the antenna heights (h_J is the height of the jamming antenna and h_S is the height of the desired transmitter antenna).

5.8.3 Other Losses

Although the spreading loss is the dominant factor, and J/S equations are usually expressed this way, the jamming and desired signal propagation path also have atmospheric losses and can be subject to nonline-of-sight or rain losses. If there are large differences between the two distances or line-of-sight conditions, these calculations should be made, and the J/S ratio adjusted accordingly.

5.8.4 Digital Versus Analog Signals

When jamming analog modulated communication signals, it is normally necessary to achieve a high J/S ratio. This is necessary because a receiver operator has a significant ability to listen "adaptively." In analog voice or visually presented communication, we can "fill in the blanks" in a low-quality transmission from the context. This is particularly true in tactical military communication where important information is sent in fairly rigid formats. Examples are the old five-paragraph operations order and the phonetic alphabet.

When jamming digitally modulated communications signals, we attack the signal by trying to make it unreadable to the digital demodulator. We can either interfere with the synchronization or produce bit errors. Since the synchronization tends to be quite robust, the basic approach is to create bit errors.

You will note from Figure 5.20 that the bit-error curve approaches 50% as the signal quality is reduced. In general, it is assumed that the received signal quality is not further reduced by J/S values greater than unity (i.e., 0 dB).

Further, if the signal is (in the short term) unreadable a third of the time, it is considered useless. An example of how this occurs is when a frequency-hopping radio finds a third of the channels to which it hops occupied by strong signals.

This means that a digital signal need only be jammed to a J/S of 0 dB one-third of the time, while an analog signal requires positive J/S 100% of the time. In Section 5.9.1, we will discuss how this leads us to perform "partial band jamming" and the effect of error-correction codes on jamming effectiveness.

5.9 Jamming Spread Spectrum Signals

Spread spectrum signals are subject to the same jamming equations as any other signals, but the ability of the cooperative receiver to collapse the spectrum gives it a "processing gain" that reduces the effectiveness of the jamming. In general, the processing gain advantage is the same as the spreading ratio (i.e., the transmission bandwidth/the information bandwidth). It is also defined (in DSSS signals) as the code rate (used for spreading) divided by the data rate. Another applicable term is "jamming margin," which is defined by the following formula.

$$M_J = G_P - L_{SYS} - SNR_{OUT}$$

where

M_J = the jamming margin (in decibels);

G_P = the processing gain (in decibels);

L_{SYS} = the system losses (in decibels);

SNR_{OUT} = the required output SNR.

It is important to remember that spread spectrum signals almost always carry their information in digital form. Thus, the considerations presented in Section 5.8.4 apply to any effort to jam spread spectrum signals of any kind. This enables some techniques which help overcome the jamming protection afforded by spectrum spreading.

5.9.1 Jamming Frequency Hop Signals

If a narrow-bandwidth jamming signal is applied to a frequency-hopping receiver, it will be received only when the receiver happens to hop to that frequency. This will cause the jamming effectiveness to be significantly reduced.

For example, if a CW jamming signal is applied to a Jaguar V receiver (which hops randomly over a maximum of 2,320 channels), the receiver will see that jamming signal only 0.043% of the time. Alternately, if the jamming signal is spread over the whole 2,320 channel frequencies, the J/S per channel will be reduced by 33.65 dB. Thus, more sophisticated jammers are required for frequency-hopping signals.

5.9.1.1 Follower Jamming

The equations given in Section 5.8 for communication jamming apply directly to slow hopping FH signals. However, since the FH signal remains at one frequency only for one hop period, it is necessary for the jamming system to determine the transmission frequency and apply jamming to the desired receiver for enough of the hop period to prevent successful communication. As discussed above, digital signals need only be jammed 33% of the time with a 0-dB J/S ratio. This means that jamming FH signals requires quite moderate power, if the jammer has a receiving and processing capability which can detect the FH signal and set on a jammer in less than 57% of the hop period (i.e., 67% of the remaining hop period after the 15% synthesizer settling time). This timing is shown in Figure 5.30.

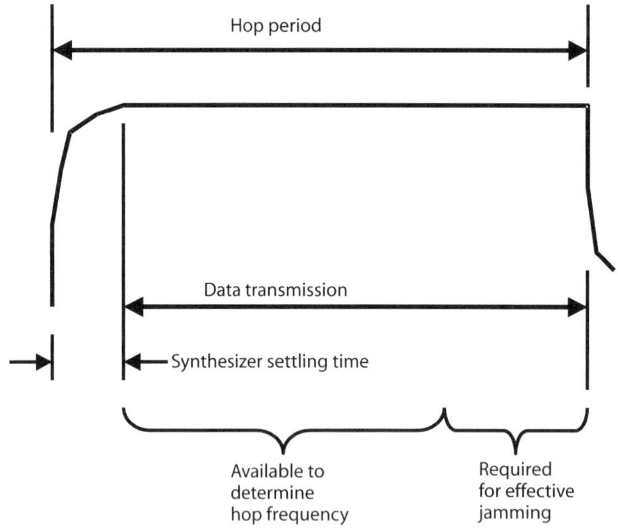

Figure 5.30 A follower jammer must determine the frequency of the hop and set its jamming frequency during 67% of the data transmission time.

A jammer that jams each hop is called a follower jammer. The necessary receiving and processing subsystem is quite complex, but is well within the state of the art of current digital receiver technology.

The overall block diagram of a digital receiver is shown in Figure 5.31. The RF front end allows the receiver to be tuned across an extended frequency range. The analog-to-digital converter (ADC) digitizes an intermediate frequency band, and the resulting digital data is input to a computer. Software in the computer then performs the rest of the receiver functions: filtering, demodulation, post demodulation processing, and output formatting.

In general, the computer can perform any function that could be performed by hardware circuitry. However, it is constrained to the resolution and accuracy of the output of the ADC and the capabilities (speed and memory capacity) of the computer. The "computer" can actually be a number of separate computers or data processors which perform parallel or sequential tasks.

The digital receiver can also perform functions that are very difficult or impractical to implement in hardware; for example, simultaneous amplitude and/or phase measurements at a number of frequencies or time compression of signals. Fast Fourier transform (FFT) software or processors provide a very large scale channelizer that can be quickly reconfigured to optimize signal processing.

With an FFT, the digital computer can determine the frequency of a hopping signal a few microseconds after it arrives at a new hop frequency. Of course, this requires other analysis capability to track fixed-frequency emitters so that the new hopping frequency can be identified. The jammer is then set to the frequency of the frequency hopper. An important consideration is that there are probably multiple frequency-hopping nets that can be seen by the receiver. This generally requires that the follower jammer be associated with an emitter location capability. The receiver is jammed with the frequency of the signal emitted by a transmitter in the location identified as containing a hopper (see Figure 5.32).

5.9.1.2 Partial-Band Jamming

Another approach to jamming a frequency hop signal is partial-band jamming. With this technique, it is necessary to determine the level of the desired signal in

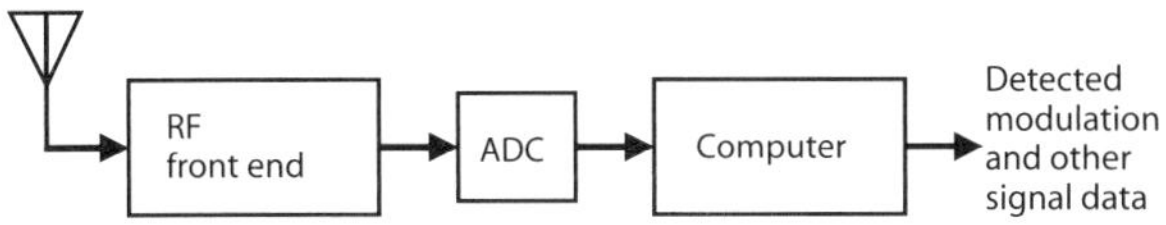

Figure 5.31 A digital receiver includes an RF front end, a digitizer, and a computer.

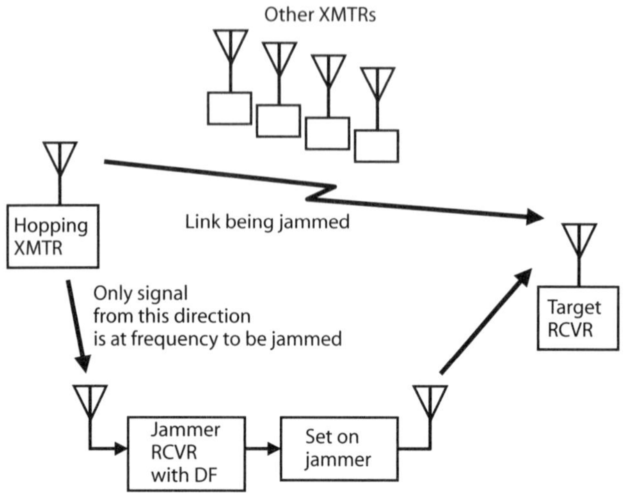

Figure 5.32 An emitter location capability is necessary for a follower jammer to determine the correct frequency on which it must jam.

the receiver. Then, the jammer power is spread over the maximum frequency range that will allow jammer power in the receiver to equal the desired signal power at each hop frequency.

As shown in Figure 5.33, if the location of the desired signal transmitter is known, a measurement of signal strength at the jammer receiver will allow calculation of the effective radiated power. In this example, it is assumed that the transmitter has a whip antenna or some other antenna type with 360° azimuth coverage. The problem is more complex when the jammed link uses directional antennas, but can still be solved. The ERP of the desired signal transmitter is the received signal strength (adjusted for the jammer receiving antenna gain) increased by the spreading loss by the formula:

$$ERP_S = P_{RJ} - G_{RJ} + 32 + 20 \log F + 20 \log d_{TJ}$$

where

ERP_S is the effective radiated power of the desired signal transmitter in dBm;

P_{RJ} is the received power in the jammer receiver in decibels;

G_{RJ} is the gain of the jammer receiving antenna in decibels;

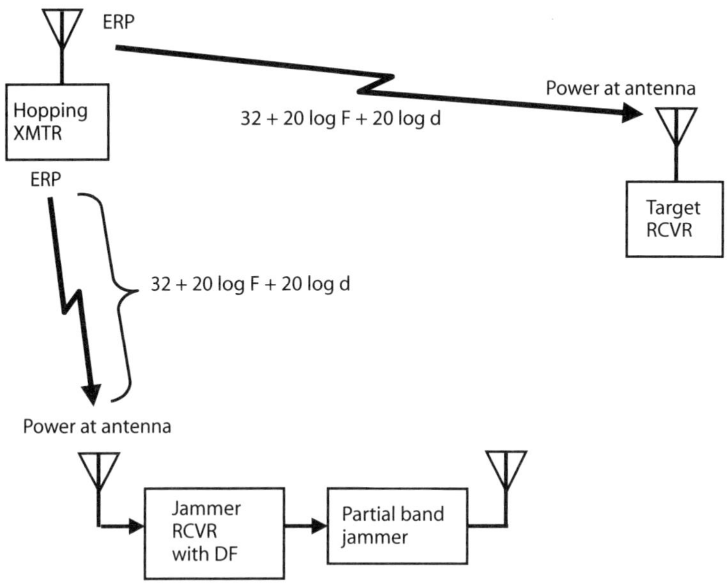

Figure 5.33 The jammer receiver can determine the transmitter ERP from received signal power if the transmitter location is known.

> F is the desired signal frequency (in megahertz);
>
> d_{TJ} is the distance from the desired signal transmitter to the jammer.

The calculated ERP_S can be adjusted for atmospheric loss and any known physical conditions such as rain and transmitter antenna directivity if appropriate; however, the spreading loss will be the primary loss factor.

If the communication link being jammed uses transceivers, the location of the receiver being jammed can be determined. With the location of the receiver, it is possible to calculate the distance from the desired transmitter to the target receiver. Then, the received desired signal power can be calculated using the spreading loss formula ($L_S = 32 + 20 \log F + 20 \log d$).

As shown in Figure 5.34, the required jammer ERP to achieve a J/S of 0 dB (i.e., received jammer power = received desired signal power) at any frequency in the hop range can be calculated from the jammer to receiver distance.

Then the total output power of the jammer transmitter is spread over the maximum frequency band that will achieve this power at the frequency of each hop as shown in Figure 5.35. For example, if the total jammer power arriving at the receiver is 20-dB higher than the desired signal power (i.e., factor of 100), the jammer could be spread over 100 channels. This would provide 0-dB J/S in each jammed channel.

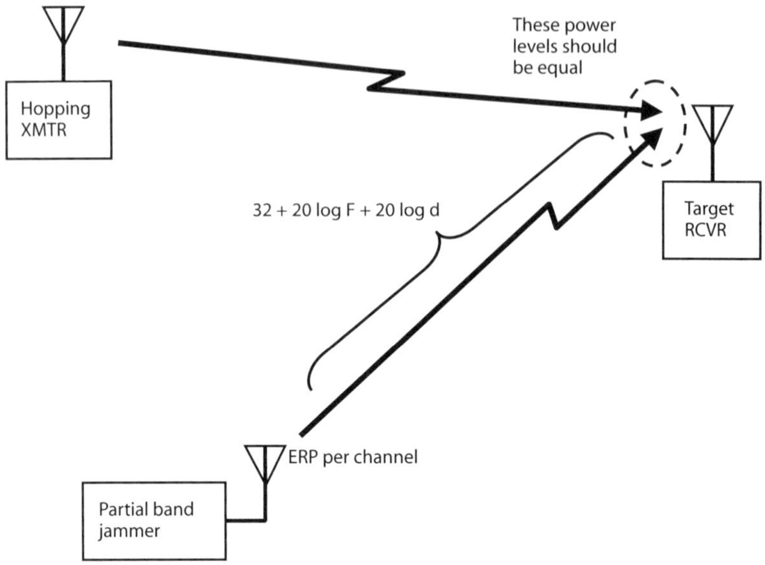

Figure 5.34 The jammer ERP per channel should provide a signal at the target receiver antenna that is equal to the power of the arriving desired signal.

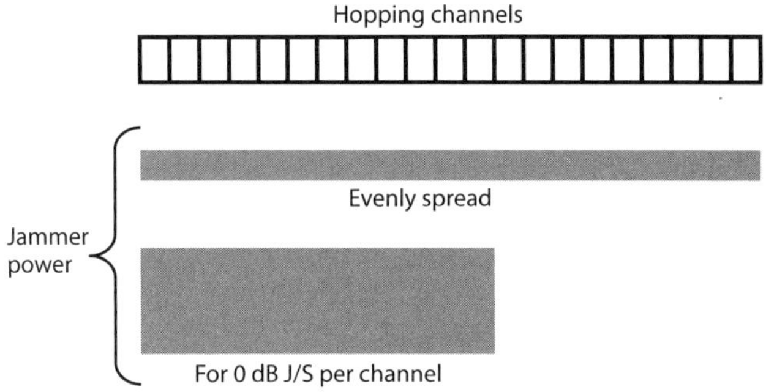

Figure 5.35 A partial-band jammer distributes its available power to achieve 0 dB J/S in each jammed channel at the jammed receiver.

If the jammer can cover one-third of the channels over which the desired signal hops, the jamming is considered effective. Even if the jammer does not cover that many channels, partial-band jamming still optimizes the jamming effectiveness.

Note that a partial-band jammer does not require a sophisticated receiver capable of very quickly detecting the frequency of each hop.

5.9.2 Jamming Chirp Signals

A narrow-bandwidth jamming signal applied to a chirp receiver will only remain in the bandwidth for a small percentage of the time, greatly reducing its jamming effectiveness.

If the tuning slope of the chirp frequency modulation can be determined and replicated, and if the start time of each frequency sweep can be detected, a follower jammer can be applied to chirp signals. Otherwise, a partial-band jammer can be used—as described above for frequency-hopping signals.

5.9.3 Jamming DSSS Signals

The direct sequence spread spectrum demodulator in the receiver is functionally exactly like the modulator in the transmitter. It performs a binary addition of the spread spectrum signal with a pseudorandom code that is synchronized with the code applied in the transmitter. Thus, if a narrow-bandwidth jamming signal is applied to the DSSS demodulator, the spectrum of that signal will be spread in the same way the desired signal was spread in the transmitter. This will reduce the detectability of the jamming signal by a factor equal to the processing gain—that is, by the spreading factor. This is shown in Figure 5.36.

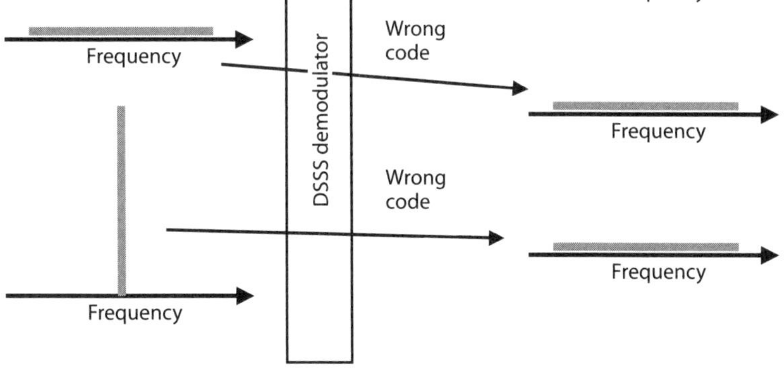

Figure 5.36 A direct sequence demodulator spreads a narrowband signal just as the DS modulator in the transmitter does.

On the other hand, if a spread spectrum signal with a nonsynchronized code is used as a jammer, that signal will not be despread, but will be down by a factor equal to the processing gain as compared with the despread desired signal (see Figure 5.36). Note that this is exactly what is done in code division multiplexing schemes like that used in the GPS system. A fixed tuned receiver looks at all GPS satellite signals, each of which has a different code sequence. The receiver tries various code sequences (each satellite has a different code) until it finds the corresponding code—allowing it to receive the signal and determine which satellite transmission it has received.

There are various jamming techniques that can be used against DSSS signals, including stand in jamming and pulse jamming.

5.9.3.1 Stand-In Jamming

A CW jammer is quite simple, and can generate relatively large ERP relative to more complex jammers. This technique involves just increasing the jammer power by enough to overcome the processing gain in the receiver. If the signal is spread by a factor of 1,000, this requires 30-dB J/S to create an effective 0-dB J/S to effectively jam a digital transmission. This can be very difficult if the jammer is far from the receiver, because transmitted signals are attenuated by the square of the transmitter (in this case the jammer) to receiver distance. However, if the range is reduced by a factor of 31.6, 30-dB more signal reaches the receiver (see Figure 5.37).

Emplaced jammers, jamming payloads on UAVs, and artillery delivered jammers are a few of the ways to get jammers much closer to target receivers.

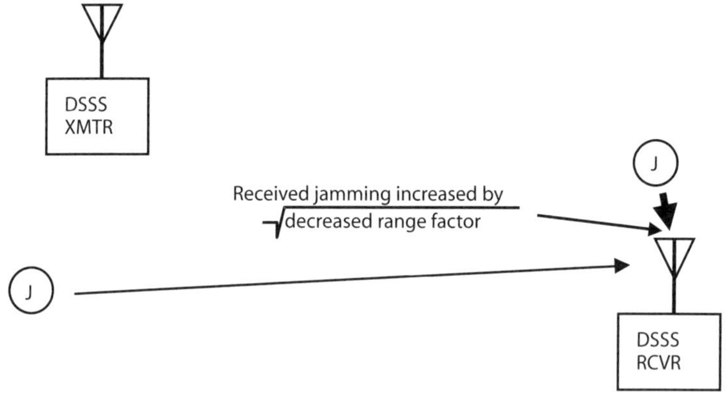

Figure 5.37 A direct sequence demodulator spreads a narrowband signal just as the DS modulator in the transmitter does.

5.9.3.2 Pulse Jamming

The peak power of a pulse can be much higher than the constant power from a continuous signal transmitter. Since we only need to jam a digital signal one-third of the time, a 33% duty factor pulse will be adequate for effective jamming. The enhanced peak power improves the J/S.

5.9.4 Impact of Error-Correction Coding

Error detection and correction (EDC) codes allow a communication system to detect errors occurring during the transmission process and to correct those errors in the receiver. There is, however, a limit to the number of errors that can be corrected by a code of any particular power. An EDC code adds bits (or bytes) to a transmitted signal in a systematic way. The more bits or bytes that are added, the higher the power of the code. Then, at the receiver, the whole digital signal (data and code bits) is processed to identify errors, which can then be corrected.

The power of the code determines the percentage of incorrect bits or bytes that can be corrected. For example, the 31,15 Reed/Solomon code that is used in JTDS is a block code. It corrects whole bytes (whether one or all bits in the byte are wrong). This code more than doubles the transmitted bit rate, but can correct 8 bad bytes out of every 31 bytes sent. Thus, as long as the received byte-error rate is less than 25.8%, the output byte error rate is zero.

This assumes, of course, that the errors are randomly spread—which may not always be the case. Consider a frequency hopper that hops to a frequency occupied by a strong single channel signal. It will have 100% bad bits during the time it is on that hop frequency. To fix this problem, the digital data is interleaved as shown in Figure 5.38. This process is designed so that no more than 8 bytes of each 31 will be transmitted on a single hop.

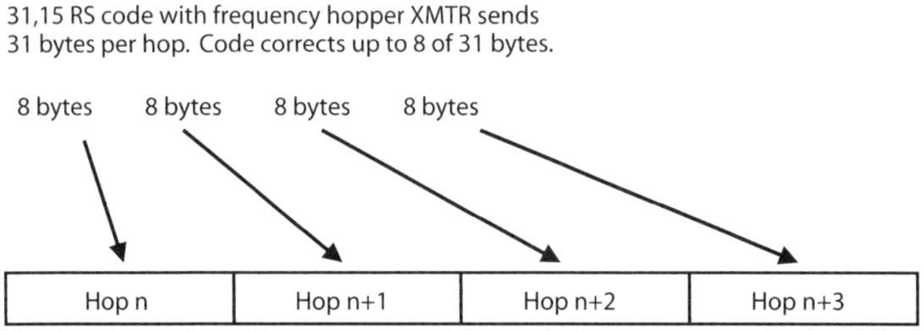

Figure 5.38 In order to allow an error-correcting code to correct all of the bits in an occupied hop, the digital data is interleaved to reduce the number of sequential bytes in a hop.

If partial-band jamming is used against a frequency hopper, some of the hop frequencies are jammed and others are not. The receiver can correct all of the byte errors on jammed hops—up to some limit. This reduces the effectiveness of the jamming and requires that more hops be jammed to adequately jam the transmission.

5.10 Location of Spread Spectrum Transmitters

In general, any of the emitter location approaches that will be described in Section 6.2 can be used to locate spread spectrum transmitters. There are, however, some special considerations for each of the three spread spectrum techniques. Some aspects of the precision emitter location techniques that will be described in Section 6.3 make their application to spread spectrum signals quite challenging.

5.10.1 Location of Frequency-Hopping Transmitters

A frequency hopper is the least challenging of the spread spectrum transmitters to locate, because it places all of its radiated power at one frequency for a period of milliseconds. The challenge is to determine that frequency before the transmitter hops to a different frequency. There are two basic approaches to this problem. One is to use a simple sweeping receiver/direction finder to determine the direction of arrival of the signal at a few of its hops during a transmission. The second is to use a very fast digital receiver/direction finder to determine the direction of arrival during every hop.

5.10.1.1 Sweeping Receiver Approach

This technique is widely used in moderately priced emitter location systems. Each direction finding station has a receiver with the general block diagram shown in Figure 5.39. The receiver usually sweeps at a high rate, pausing only long enough at each step to determine whether or not a signal is present. If there is signal power at a frequency, the receiver stops long enough to perform a DF on that signal.

Data is collected in a computer file and is sometimes presented to an operator on a frequency versus angle-of-arrival display like that shown in Figure 5.40. Each dot on this display represents a received signal. Note that there are several intercepts at the same angle but at different frequencies. This is the characteristic of a frequency hopper. If a number of hits at the same angle of arrival are detected during a signal transmission (generally specified at a small number of seconds), a frequency hopper direction of arrival is reported. This same process is repeated at a second (and preferably a third) DF station so that triangulation can be performed to locate the transmitter.

As mentioned in the introduction to this chapter, the density of communication signals in a tactical military environment can be very high. Numbers such as 10% channel occupancy are often stated as assumptions in system specifications. This means that at any instant, 10% of the channels are occupied. If you integrate data for a few minutes, they will be closer to 100% occupied. Multiple frequency-hopping radios can be expected to operate in a signal net, and multiple, independent nets can be expected. This complicates the triangulation

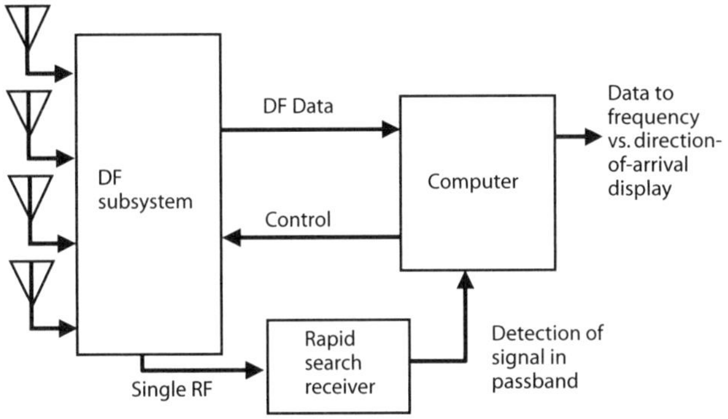

Figure 5.39 A sweeping DF system for frequency hoppers includes a rapid search receiver to detect occupied channels. Then the search is stopped while a DF is taken.

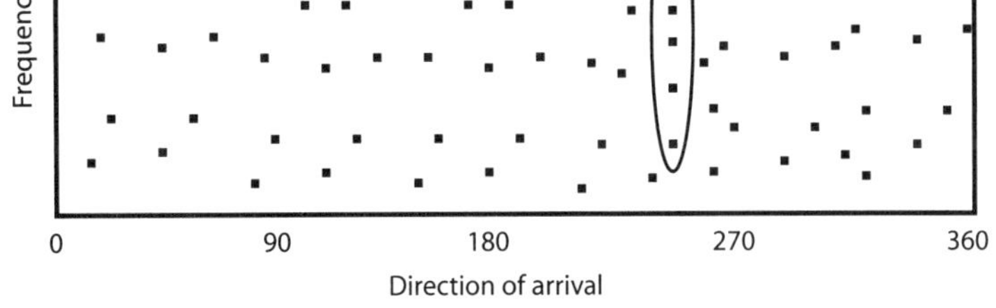

Figure 5.40 When collected DOA data shows multiple frequencies at one angle of arrival, a frequency hopper is identified.

problem. Consider a very simple environment with two hopping emitters being measured by two direction-finding systems as shown in Figure 5.41. There are four possible emitter locations in this simple case, and in the real world it is a lot worse.

The solution is to cause the two systems to step together. Each frequency hopper can be at any frequency in its range at any given step, but if the two receivers are locked together, they will always look at the same frequency and will thus catch the same emitter on the same hop.

5.10.1.2 High-Speed Digital Receiver Approach

The digital receiver discussed in Section 5.9.1 can be implemented to work as a direction finder. With two digital receivers, each connected to a different antenna, the received power in each FFT channel allows amplitude comparison direction finding to be implemented to determine the angle-of-arrival of all signals in the environment. Signals which hop to different frequencies but have the same direction of arrival are identified as hoppers. Two such direction-finding stations can be netted to triangulate the emitter locations of all fixed frequency and hopping emitters in the environment.

If there are two FFT processors in parallel in each receiver, one operating on a quarter-wavelength delayed input signal, the receiver can generate I&Q samples in each FFT channel. This preserves the receive phase of signals in each channel, allowing the development of a multiple channel interferometric direction finding system.

The resulting emitter location data can support follower jamming as well as normal order of battle reporting requirements in an environment containing frequency hoppers.

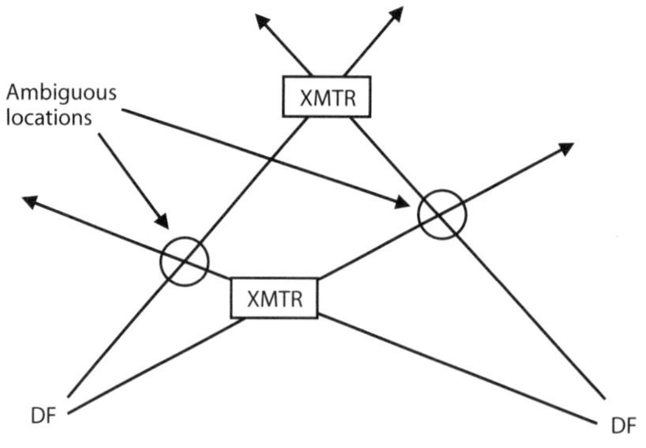

Figure 5.41 If two frequency-hopping emitters are found by two noncorrelated direction-finding systems, there will be four possible emitter locations.

5.10.2 Location of Chirped Transmitters

If receivers with wide enough bandwidth to collect a reasonable portion of the frequency sweep are used in an amplitude comparison DF system, the direction of arrival for chirp signals can be measured. This has been done successfully with a Watson-Watt direction-finding system as described in Section 6.2.3.

5.10.3 Location of Direct Sequence Transmitters

DSSS transmitters can be located with chip detection and with energy detection techniques.

The chips (the digital bits used for spreading the signal) necessarily have highly predictable transition times. Using hardware or software to create tapped delay lines at the chip rate allows this chip transition energy to be integrated. This allows the use of amplitude comparison techniques to measure direction of arrival. Two such systems can triangulate the emitter location.

There are also energy detection techniques, extensively discussed in an excellent book by Robin and George Dillard, *Detectability of Spread Spectrum Signals.* The detected energy levels allow the use of narrow-beam antennas for direction-of-arrival measurement, and could be used to implement a multiple antenna amplitude comparison direction finding system.

5.10.4 Precision Emitter Location Techniques

The location of spread spectrum emitters using the precision emitter location techniques as described in Section 6.3 is very difficult because communication signals use continuous modulations such as AM and FM, and thus require significant correlation times that are much greater than hop times. The pseudorandom parameters of chirp and DSSS signals also make correlation very difficult to achieve.

6

Accuracy of Emitter Location Systems

Almost every military asset, from large fixed installations to an individual aircraft, vehicle, or small unit must transmit signals of one kind or another in order to perform its mission. Although the asset may not be seen at night, in the presence of fog or smoke, when camouflaged, or when beyond line of sight; the location of its transmitters correspond to its physical location. Analysis of the signals transmitted from that location can usually identify the type of asset (weapon, military unit, aircraft, ship). Location and identification of the asset support the following military activities:

- *Warning of imminent attack.* By locating a hostile weapon platform (aircraft, ship, artillery unit) that is within engagement range, you can determine what kind of attack will likely take place, and the direction from which the weapon will approach us.
- *Threat avoidance.* If you know where the threat systems associated with the transmitters are located, you can avoid those areas or at least go into those areas forewarned.
- *Selection and implementation of electronic countermeasures.* The location and identification of a threat will determine the types of countermeasures that will be effective and when they should be initiated.
- *Development of electronic order of battle.* Since different types of units have different types of transmitters (with different frequencies and modulations) the emitters can, to some degree of accuracy, identify the types of enemy units. By knowing the location and identification of

enemy units and their recent movement history, it is possible to estimate the enemy's force structure and even predict its actions. (For example, it takes certain types and numbers of units to attack versus to hold a defensive position.)

- *Targeting.* With very accurate emitter location information, it may be practical to bring weapons to bear on important enemy assets beyond visual range. Identification of assets by their associated emitters can also enable more accurate target identification than can be achieved by optical techniques alone.
- *Cueing of narrow field of view reconnaissance assets.* Optical sensors typically have a significant trade-off of field of view against resolution. Therefore, they are often described as "soda straw" type sensors (i.e., like looking at something through a soda straw). The point is that it may take a long time to find an enemy asset by scanning this narrow field of view. An electronic emitter location system can reduce the search area by providing an approximate target location.

This chapter starts with a review of emitter location techniques, but just to the depth required to support the primary chapter focus on the way the accuracy of emitter location systems is specified and measured. A more thorough discussion of emitter location systems is provided in Chapter 8 of *EW 101*.

Accuracies are typically stated in terms of the RMS error of the system; a calculated value that is widely accepted as defining the effective accuracy of the system. The RMS error is determined by gathering a large amount of angular- or location-error data—at many frequencies, at many angles, or both. Each datum value is squared, and the square root of the mean of the squared values is calculated. (There will be more on RMS error in Section 6.4.1.)

6.1 Basic Emitter Location Approaches

As shown in Figure 6.1, there are three basic ways in which emitters are located. This figure shows techniques for location of emitters in two dimensions, for example on the Earth's surface. All approaches can, of course, be expanded to three dimensions. The first approach [Figure 6.1(a)] is triangulation. That is by the determination of lines of bearing from two (or more) known locations to the emitter. The point at which these lines of bearing cross is the emitter location. The second [Figure 6.1(b)] is determination of a distance and one line of bearing. The third approach [Figure 6.1(c)] is used in precision location systems. It

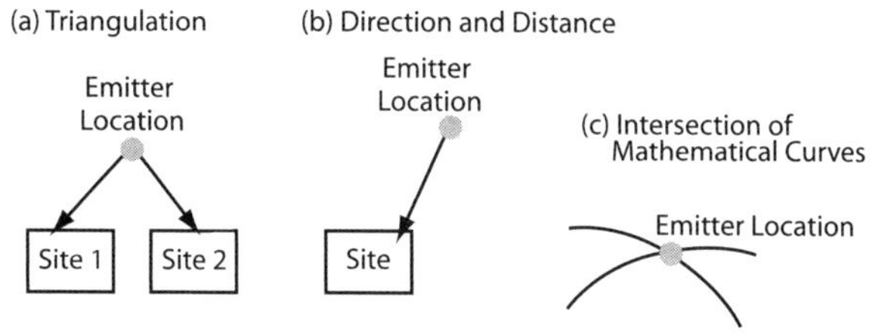

Figure 6.1 Three basic approaches to emitter location are: (a) triangulation, (b) angle and distance measurement, and (c) intersection of mathematically derived curves.

involves the determination of two mathematically described curves that cross at the emitter location.

While it is also possible to locate an emitter by determining the distance from two or more known locations, there are practical considerations that normally limit this approach to the location of cooperative emitters. Since electronic warfare systems locate noncooperative enemy emitters, we will ignore this approach at this time.

6.2 Angle Measurement Techniques

In the first two approaches, it is necessary to determine a line of bearing from known locations to the emitter location. This is accomplished by measuring the direction-of-arrival (DOA) of signals. It is often called direction finding (DF). The principal direction-finding techniques are:

- Rotating directional antenna;
- Multiple antenna amplitude comparison;
- Watson-Watt;
- Doppler;
- Interferometer.

6.2.1 Rotating Directional Antenna

As shown in Figure 6.2, a rotating antenna has a gain pattern that is a function of the angle from the antenna boresight. By rotating the antenna past an emitter, the direction of arrival of the signal can be determined from the time history of the antenna's orientation. It is important to note that the shape of the antenna gain curve is well documented, so two or more intercepts within the

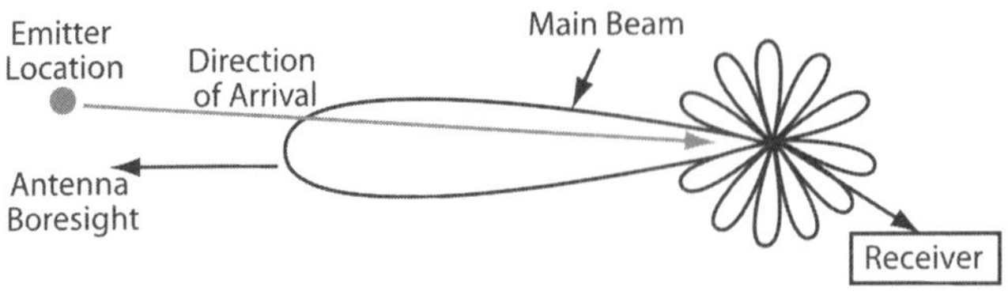

Figure 6.2 The gain pattern of the receiving antenna varies with angle from the boresight.

main beam are sufficient to determine the orientation which would place the signal at the antenna boresight. Large antennas with narrow beams allow very good direction-of-arrival measurement accuracy (of the order of one-tenth of a beamwidth). This approach is very common in naval EW systems in which relatively large antennas are practical.

6.2.2 Multiple Antenna Amplitude Comparison

As shown in Figure 6.3, the gain patterns of two differently oriented antennas provide an output signal amplitude ratio when the same signal is intercepted by both. The direction of arrival of the signal can be calculated from the amplitude ratio. This technique is widely used in radar-warning receivers on aircraft and smaller naval craft, because it does not require large antennas and is fast enough to determine the direction of arrival of single pulses. Its accuracy is typically low (on the order of 5° to 15°).

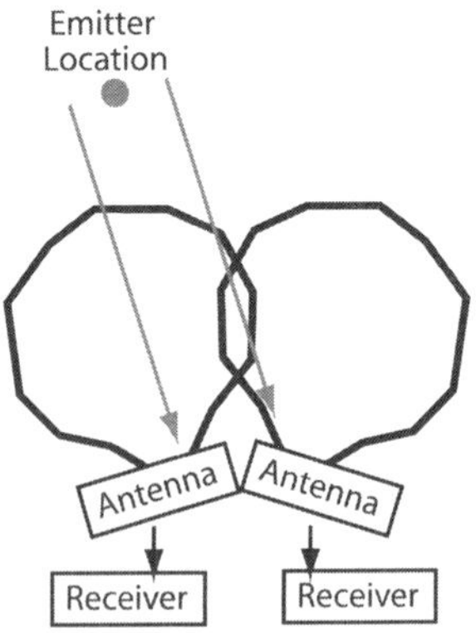

Figure 6.3 The gain pattern of the two antennas provides a power ratio to the two receivers.

6.2.3 Watson-Watt Technique

The Watson-Watt technique, developed by Sir Robert Watson-Watt (of radar development fame) uses three antennas in a line. The center antenna is a sense antenna and the two outer antennas are approximately a quarter-wavelength apart. The sum and difference of the signals from the outer antennas (normalized by the sense antenna) create a cardioid pattern versus angle as shown in Figure 6.4. If there are several symmetrical pairs of outside antennas, switching among the pairs causes the cardioid pattern to rotate, allowing calculation of the direction of arrival of the signal. This technique is widely used in emitter location systems proving moderate DOA accuracy (about 2.5° RMS).

6.2.4 Doppler Technique

This technique measures the frequency of a received signal in two antennas, one of which is rotating around the other as shown in Figure 6.5. The frequency of the signal received by the moving antenna is different from the transmitted frequency because the moving antenna has a Doppler shift proportional to the rate of change of transmission distance. The nonmoving antenna receives the transmitted frequency. The circular motion of the moving antenna (A) causes a sinusoidal Doppler shift relative to the frequency received by antenna B. The direction of arrival of the emitter is the angle at which the Doppler shift goes from positive to negative. In practical systems, the rotating antenna is replaced

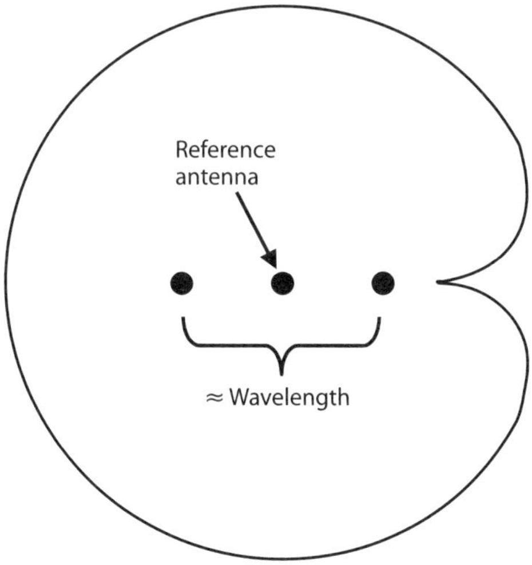

Figure 6.4 In the Watson-Watt DF system, the sum and difference patterns of the two outside antennas generate a cardioid pattern.

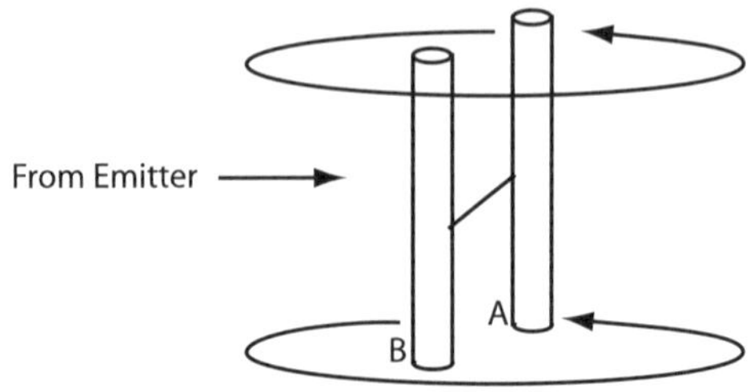

Figure 6.5 In the Doppler DF system, antenna A rotates around antenna B, generating a sinusoidal frequency offset from the frequency received by antenna B.

by a circular array of antennas sequentially switched into a receiver. Comparison of the signal phase at adjacent antennas as they are switched allows the calculation of the frequency. This is widely used in radio DF systems on civilian watercraft. It typically yields 3° or more RMS accuracy.

6.2.5 Distance Measuring Techniques

If both the transmitted and received power levels are known, it is possible to calculate the distance over which a signal has been transmitted—as shown in Figure 6.6. Since this technique is used only in EW systems which do not require high-accuracy distance measurement, it is common practice to ignore all factors except for the spreading (or space) loss. The spreading loss is given by the formula:

$$L_S = 32.4 + 20\log(F) + 20\log(d)$$

where

L_S is the spreading loss in decibels;

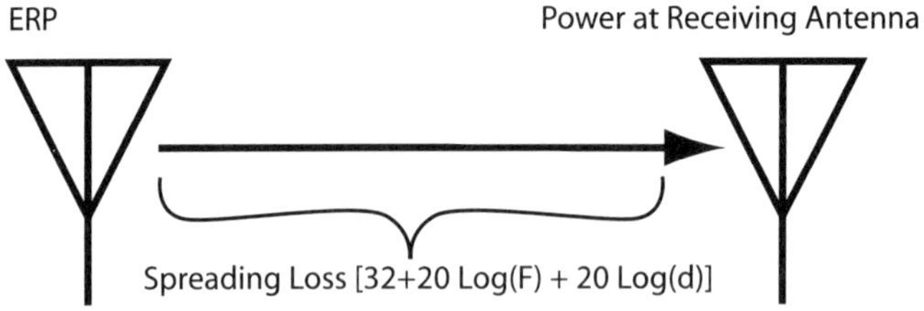

Figure 6.6 The power at the receiving antenna is reduced from the ERP by a factor of frequency and distance.

F is the transmitted frequency in megahertz;

d is the transmission path length in kilometers.

This equation can be solved for d.

$$d = \text{antilog}\{[L_S - 32.4 - 20\log(F)]/20\}$$

$$\text{antilog}(fn) \text{ is actually } 10^{fn}$$

For example, if a radar operating at 10 GHz is known to have an effective radiated power of +100 dBm and arrives at the receiving antenna at –50 dBm, the spreading loss is 150 dB. Plugging the values into the equation yields a distance of approximately 76 km.

In practical systems, particularly on aircraft, the accuracy of this measurement may be no better than 25% of the measured range.

A much more accurate technique involves the measuring propagation time. Signals travel at very close to the speed of light (3×10^8 m/s). This is very close to 1 ft/ns. Thus, if the time a signal leaves the transmit antenna and the time it arrives at the receiving antenna are known, the exact propagation distance can be determined from the formula

$$d = tc$$

where

d = the propagation distance in meters;

t = the transmission time in seconds;

c = the speed of light (3×10^8 m/s).

For example, if the transmission time is 1 ms, the distance is 300 km.

This is the way radars measure distance—easy because the transmitter and receiver are typically colocated. However, it is much harder in one-way communication. The problem is the accurate determination of the time of transmission and the time of arrival. The time-of-arrival problem has been largely solved by use of very accurate GPS-based clocks, but the time of transmission can only be determined in cooperative systems (such as GPS).

As will be discussed is Section 6.3, one of the important precision emitter location techniques (for hostile emitters) is based on measurement of the difference between the times of arrival of a signal at two receiving stations.

6.2.6 Interferometric Direction Finding

When a direction-finding system is specified at about 1° RMS accuracy, it usually uses the interferometric technique. This technique measures the phase of a received signal at each of two antennas and derives the direction of arrival of the signal from the difference between those two phase values. The basis of the interferometric direction-finding technique is best explained in terms of the interferometric triangle in Figure 6.7.

The two antennas form a baseline. It is assumed that the system knows the locations of these two antennas, so their separation and orientation can be accurately calculated. Now consider the "wave front" of the arriving signal. The wave front does not exist in nature, but it is a useful concept. This is a line perpendicular to the direction from which the signal arrives at the receiving system location.

Consider that a transmitted signal is sinusoidal and that it propagates at the speed of light. The length of a full cycle of the signal (the wavelength) includes 360° of phase. The observed phase of the signal would be the same at any point along the wave front. Thus, the wave front can be considered a line of constant phase (for example, the positive going zero crossing). The relationship between wavelength and frequency is given by the formula:

$$c = \lambda F$$

where

c = the speed of light (3×10^8 m/s);

λ = the wavelength in meters;

F = the signal frequency in hertz.

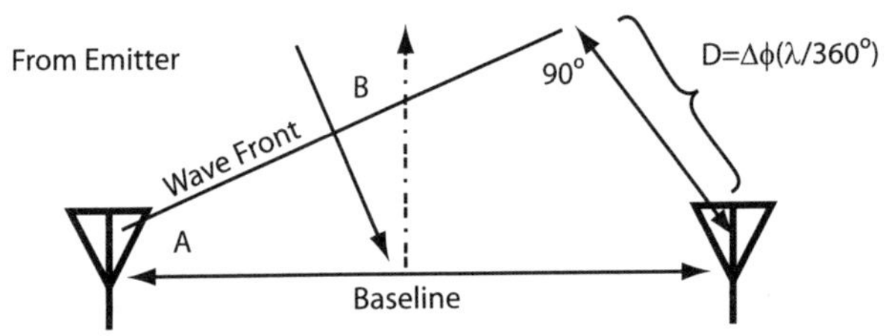

Figure 6.7 The interferometric triangle shows the geometry of the baseline formed by two antennas and the way that the angle of arrival is determined from the phase difference of the signals at the two antennas.

In the interferometric triangle, the wave front touches one of the antennas and is a distance *D* from the other antenna. By construction, this is a right triangle formed by the baseline, the wave front and *D*. The ratio of *D* to the wavelength is the same as the number of degrees of phase divided by 360. The phase difference between the signal received at the two antennas is thus the ratio of *D* to a wavelength (which can, of course, be calculated from the measured frequency of the received signal).

The ratio of *D* to the baseline length is the sine of the angle *A*, and angle *A* is equal to angle *B*. Since the perpendicular to the baseline is considered the "zero" angle of an interferometer, angle *B* is the measured angle of arrival. The angle of arrival at the site also includes the orientation of the baseline.

In "single baseline" interferometer systems (i.e., one baseline used at a time), the baseline is between 0.1 and 0.5λ. A baseline less than 0.1λ does not provide adequate accuracy, and if over 0.5λ, it produces ambiguous answers.

There is also the so-called front/back ambiguity if the antennas have 360° coverage. A signal arriving from the mirror image direction would cause the same phase difference between the antennas. This is resolved by use of antennas with high front-to-back ratio or by use of multiple baselines. Figure 6.8 shows a four-dipole array used in many interferometer systems. From the top view, you can see there are six pairs of antennas (i.e., six baselines) in this array. The correct angle of arrival will correlate between data from different baselines.

There are also correlative interferometers which use baselines greater than 0.5λ, which correlate data from several baselines to resolve ambiguities. Multiple baseline precision interferometers collect simultaneous data from three or more baselines that are multiple half-wavelengths long with two baselines different in length by 0.5λ. They are able to mathematically resolve the very high number of ambiguities.

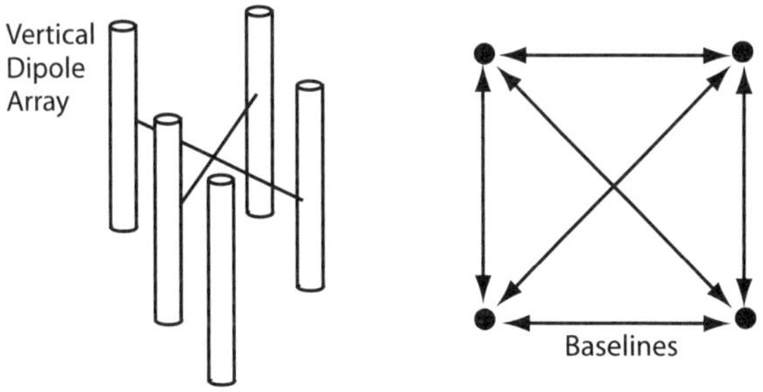

Figure 6.8 Four vertical dipoles in an array are often used in interferometer DF system to form six baselines.

6.3 Precision Emitter Location Techniques

Precision emitter location techniques are generally those considered accurate enough for targeting, providing location accuracy in tens of meters. Two techniques are typically used for precision emitter location: time difference of arrival (TDOA) and frequency difference of arrival (FDOA). They are often used together, and normally used in conjunction with less accurate location systems. First, we will focus on the TDOA approach.

6.3.1 Time Difference of Arrival Method

Section 6.2.5 discussed the calculation of propagation distance from the signal transition time. The signal travels at the speed of light, so if we know when the signal left the transmitter and arrived at the receiver, we know the path length. When dealing with cooperative signals (such as GPS) or our own data link, coding on the signal can allow the determination of the time of departure. However, when dealing with hostile emitters, we have no way of knowing when the signal left the transmitter. The only information we can measure is when the signal arrives. However, by determining the difference between the times of arrival at two sites, we can know that the transmitting site is located along a hyperbolic curve. If the time difference of arrival is measured very accurately, the emitter location will be very close to the line, but since a hyperbola is an infinite curve, the location problem is not yet solved.

Figure 6.9 shows two sites receiving signals from a single transmitter. The two sites form a baseline. The area of uncertainty is the area which might contain the emitter of interest. Note that the difference between the two distances

Figure 6.9 Two receiver sites form a baseline against which the relative distances to an emitter can be calculated from the differences in the propagation times.

determine the time-of-arrival difference. Figure 6.10 shows a few of the infinite number of hyperbolas. Each curve represents a specific time-of-arrival difference, and is called an "isochrone."

6.3.1.1 Pulse Emitter Location

The determination of the time difference of arrival is fairly easy if the transmitted signal is a pulse—given that there is a very accurate clock at each receiver location and that the measured times can be transmitted to a common location. As shown in Figure 6.11, the leading edge of the pulse gives a

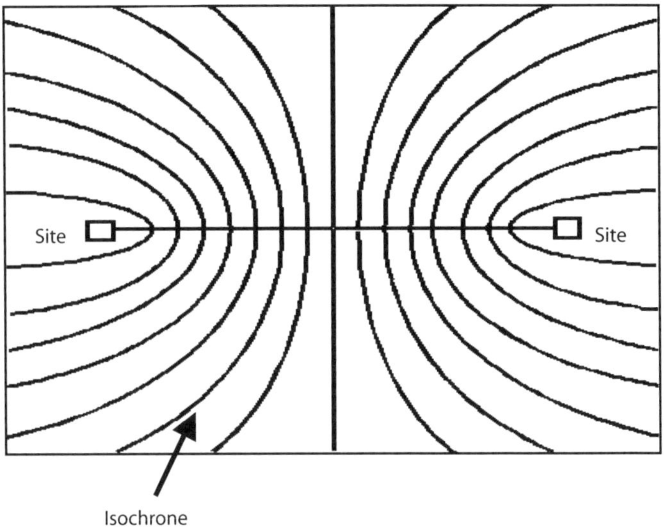

Figure 6.10 An isochrone is a hyperbolic line containing all of the locations at which an emitter could be located for a fixed difference in propagation path length to the two sites, causing a fixed time difference of arrival for a signal.

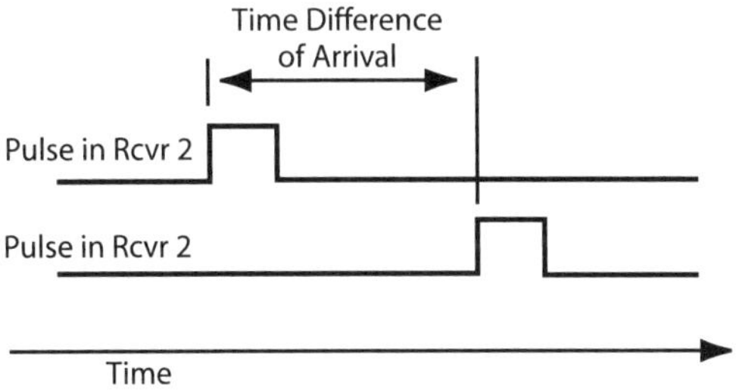

Figure 6.11 If the time of arrival of a pulse at each receiver is measured with an accurate clock, the calculation of the time difference of arrival is straightforward.

dependable time measurement event. One issue is that the two receivers must measure the same pulse. Since only a few pulses must be measured to determine the emitter location, very little data link bandwidth is required to carry the time-of-arrival data to the processing location—where the TDOA is calculated.

6.3.1.2 Analog Emitter Location

Now consider the case of a signal with analog modulation. This type of signal has a continuous carrier (at the transmit frequency) and carries its information in the modulation of the frequency, amplitude, or phase of that carrier. The carrier repeats every wavelength (typically less than a meter), so the only attribute of the signal that we can correlate to determine the time of arrival at two receivers is the modulation. The time-of-arrival difference is determined by sampling the received signal many times with a varying time delay at one of the receivers. This time delay must be varied over sufficient range to cover the minimum to the maximum time difference possible over the area in which the emitter could be located. The samples are digitized, time coded, and sent to a common point at which the correlation between the two samples can be calculated.

The correlation changes as a function of the differential delay as shown in Figure 6.12. The correlation peak occurs at the differential delay value equal to the time difference of arrival. Note that this correlation curve has a fairly smooth top, but that the peak is typically determined to the order of one-tenth of the delay increments.

The TDOA process is relatively slow for analog signals because many samples must be taken, and requires significant data transmission bandwidth because many bits per sample are required for adequate location accuracy.

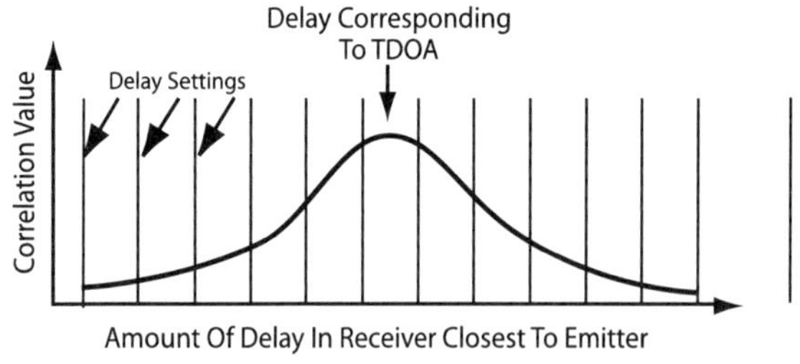

Figure 6.12 The correlation between the signals arriving at two receivers as a function of the delay of the signal in one receiver peaks at the delay value equal to the time difference of arrival. For analog signals, the curve has a soft peak.

6.3.1.3 Location

Determining the actual location of the emitter requires a third receiver site so that there can be at least two baselines. As shown in Figure 6.13, each baseline forms a hyperbolic isochrone. These two hyperbolas cross at the emitter location. There is a location ambiguity in that the two hyperbolas may cross in two locations. However, only one of these locations would be expected to lie within the area of uncertainty (shown in Figure 6.9).

In order to provide an accurate emitter location, it is necessary that the receiver site locations be accurately known. With the availability of GPS, accurate locations are available for small vehicles and even dismounted operators. If the receivers are moving, it is, of course, necessary to consider the instantaneous receiver locations when making the isochrone and emitter location calculations.

6.3.2 Precision Emitter Location by FDOA

FDOA is one of the techniques for achieving precision emitter location. It involves the measurement of the difference between the received frequency at two moving receivers from a single transmitter (which is generally not moving). Because the difference in received frequency is caused by differences in Doppler shift, FDOA is also called differential Doppler (DD).

6.3.2.1 Frequency Difference of Arrival Method

First, consider the received frequency of a signal from a fixed transmitter if the receiver is moving. As shown in Figure 6.14, the received signal frequency depends on the transmitted frequency, the speed of the receiver, and the true spherical angle between the transmitter and the velocity vector of the receiver. The received signal frequency is given by the formula

$$F_R = F_T\left\{1 + \left[V_r \cos(\theta) / \mathrm{c}\right]\right\}$$

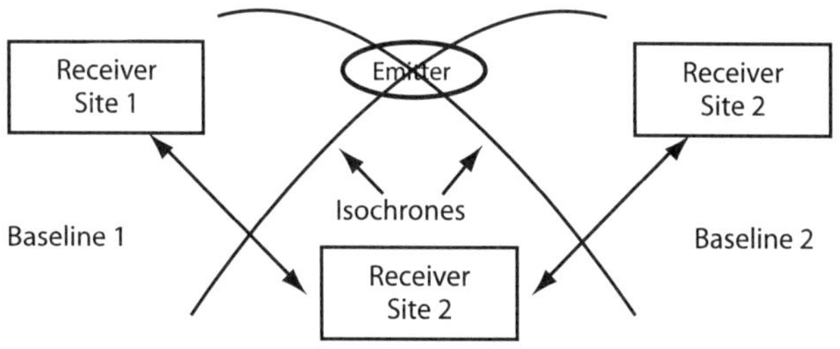

Figure 6.13 Three receiving sites allow a second baseline. The isochromes from these two baselines cross at the emitter location.

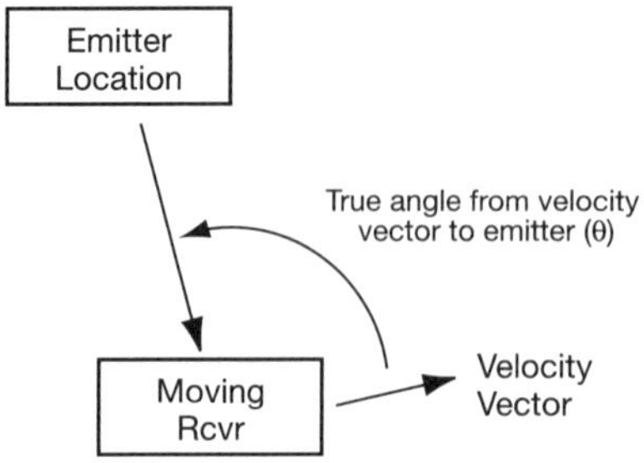

Figure 6.14 A moving receiver receives signals from a fixed emitter that are changed by a Doppler shift that is a function of velocity and angle θ.

where

F_R = the received frequency;

F_T = the transmitted frequency;

V_R = the speed of the receiver;

θ = the angle from the receiver velocity vector to the transmitter;

c = the speed of light.

Now consider two moving receivers receiving the same signal from different locations as shown in Figure 6.15. The instantaneous locations of the two receivers form a baseline. The difference between the received frequencies at the two receivers is a function of the difference between θ_1 and θ_2 and the velocity vectors of the receivers. The difference between the two received frequencies is given by the formula:

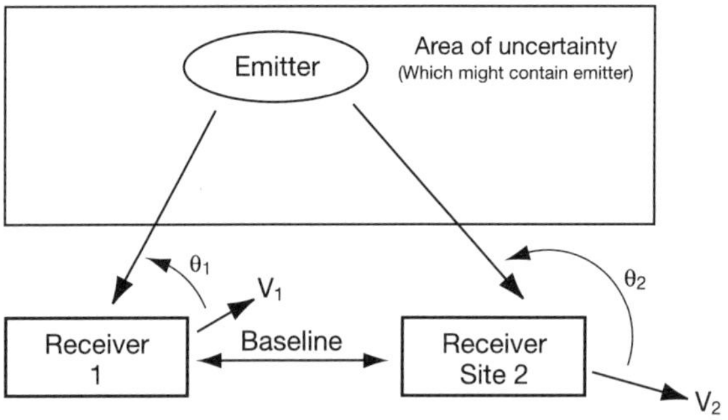

Figure 6.15 Two moving frequencies measure different received frequencies depending on their velocity vectors and intercept geometry.

$$\Delta F = F_T \left[V_2 \cos(\theta_2) - V_1 \cos(\theta_1) \right] / c$$

where

ΔF = the difference frequency;

F_T = the transmitter frequency;

V_1 = the speed of receiver 1;

V_2 = the speed of receiver 2;

θ_1 = the true spherical angle from the velocity vector of receiver 1 to the transmitter;

θ_2 = the true spherical angle from the velocity vector of receiver 2 to the transmitter;

c = the speed of light.

There is a curving three-dimensional surface defining all of the possible transmitter locations which would produce the measured frequency difference under the existing conditions. If we look at the intersection of this surface with a plane (e.g., the surface of the Earth), the resulting curve is often called an "isofreq." The two receivers can be moving in different directions at different speeds, and system computers can draw the correct isofreq curve for each velocity/geometry/frequency-difference condition. However, to simplify the visual presentation for us humans, Figure 6.16 shows a set of isofreq curves for various frequency differences for two receivers going the same direction at the same speed (but not necessarily in a tail chase). Note that this set of curves fills all space. It looks like lines of magnetic flux (as though the two receivers were the ends of a bar magnet in a high school physics book).

Like TDOA, the frequency difference measurement by two receivers does not define a location, it only defines a curve of possible positions (i.e., the isofreq). However, if the frequency difference is measured accurately, the transmitter location will be very close to the isofreq curve (order of magnitude of 50m). With a third moving receiver, we have three measurement baselines, each of which can collect FDOA data and calculate isofreqs. The location of the transmitter can then be determined from the intersection of isofreqs from two or more baselines.

6.3.3 FDOA Against Moving Transmitters

There is a significant problem associated with the FDOA location of a moving transmitter (using moving receivers). The measured frequency differences are

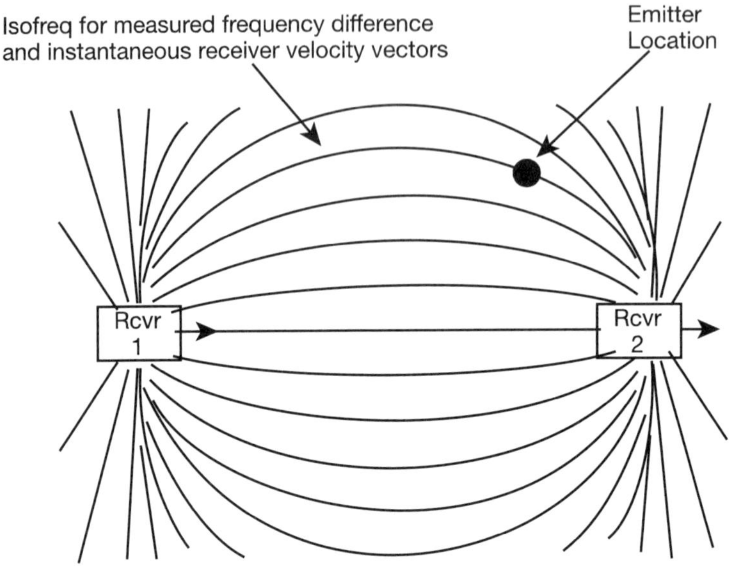

Figure 6.16 With two receivers, an FDOA system determines a curve (the isofreq) that passes through the emitter location.

from the Doppler shift caused by the accurately known velocity vectors of the two receivers. If the transmitter is also moving, it causes Doppler shifts of the same order of magnitude as those from the moving receivers—but the transmitter velocity vector is not known. This brings another variable into the emitter location calculation. While this should be mathematically resolvable, the required calculations (i.e., computer power and time required) are much more complex. Therefore, FDOA is generally accepted as being appropriate only for locating fixed or very slowly moving transmitters from moving airborne receivers.

6.3.4 Combined FDOA and TDOA

Since the measurement of frequency and time both require the presence of a highly accurate frequency reference, it is logical to perform both functions from the same two receivers. This is done in many precision location systems. Consider Figure 6.17, a set of isochrons (from TDOA) and a set of isofreqs (from FDOA) calculated for a single baseline of two receivers. You will note that the transmitter location is at the intersection of an isochron and an isofreq. Thus, a precision emitter location is determined from a single baseline of two receivers.

In practice, location systems typically use three or more platforms, so many solutions can be calculated—from TDOA or FDOA alone and by

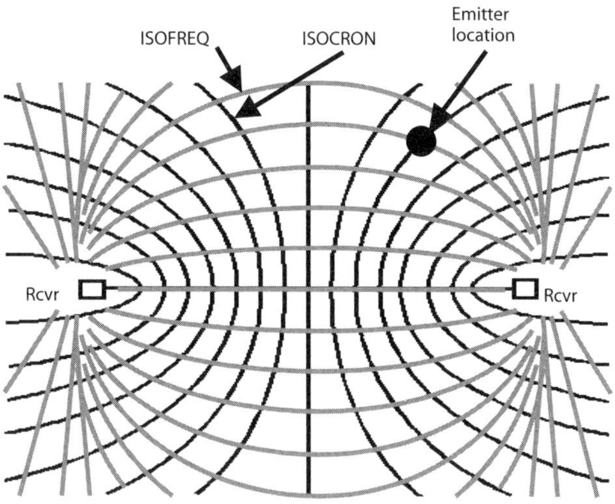

Figure 6.17 An emitter location can be determined by TDOA and FDOA from the same two moving receivers.

combined TDOA and FDOA. This multiplicity of solutions allows a more accurate ultimate result over a wide range of operational conditions.

6.4 Emitter Location–Reporting Location Accuracy

When comparing the operational value of direction-finding systems, one key parameter is the effective accuracy. For angle-of-arrival (AOA) systems, this accuracy is usually stated as the RMS angular error. In full emitter location systems (e.g., multiple angle-of-arrival DF stations), the emitter location accuracy is often stated as the CEP or the EEP.

When an emitter location is reported to someone making decisions based on that information—for example, a commander trying to determine where an enemy asset is located—the CEP or EEP defines the location uncertainty in the measurements being reported from the emitter location system. This circle or ellipse drawn on a map display can be useful in the decision-making process.

6.4.1 RMS Error

There is always some error in any direction of arrival or emitter location measurement, but we need to be able to evaluate and talk about the "effective" error in a system. If it is critical that every single measurement have less than some specific angular error, the peak error will be specified. However, in practical direction-finding systems, particularly those which instantaneously cover 360°, there can be a few specific angles and frequencies at which the measured errors are significantly larger than average. This almost always occurs in field tests, where low-level interfering signals or point reflectors can cause peak errors that have nothing to do with the performance of the system itself. If these peak errors (from inside or outside the system) occur only at a few angle/frequency combinations, they are not normally considered a fair test of the operational usefulness of the system. Therefore, we usually consider the effective accuracy of the system to be better represented by the root, mean, square (RMS) error.

To determine the RMS error, data is taken at many different angles through the full system angular range (typically 360°) and at regular increments over the system's full frequency range. Each measured angle of arrival is compared to the actual angle of arrival to determine the error angle. The actual angle is determined from the position of a turntable on which the system is parked, from the navigation system of the aircraft or ship on which the system is mounted, or from some other independent angular reference. Then, each error is squared, the squared errors averaged, and the square root of the average calculated. The formula is:

$$\text{error}_{\text{RMS}} = \sqrt{\frac{\sum_{i=1}^{n} (\text{error}_i)^2}{n}}$$

The RMS error is often considered to have two parts, the mean and the standard deviation. The mean error is simply the average of all error measurements—which can be corrected in all output data. The standard deviation is the RMS error that is calculated if the mean error has been subtracted from each data point (in effect, the RMS error without the mean error component). The relationship between the RMS error, the mean error, and the standard deviation is:

$$\text{RMS} = \sqrt{\mu^2 + \sigma^2}$$

where

μ = the mean of the data points;

σ = the standard deviation.

If you have the urge to try the formulas, you will find that the numbers 1, 4, 6, 8, and 12 yield a mean of 6.2, a standard deviation of 3.7, and an RMS of 7.22.

6.4.2 Circular Error Probable

CEP is an artillery and bombing term. If a number of rounds or bombs are aimed at an aiming stake and the distance from the stake to each hit location is measured, the circular error probable is the radius of a circle that would contain half of the hit locations.

When evaluating the performance of an emitter location system, we use CEP to quantify the location accuracy. The CEP in this case is the radius of a circle centered on the calculated emitter location that provides a 50% probability that the actual emitter location is within the circle. This is also stated as a 50% CEP or (if the circle provides a 90% chance of containing the emitter) a 90% CEP.

Consider the emitter location geometry shown in Figure 6.18. This is an ideal geometry because the two DF stations are equal distances from the emitter and 90 degrees apart when viewed from the emitter location.

Assuming that the errors in each of the DF receivers are normally distributed with zero mean error, the lines defining the RMS error values will contain 68.2% of the DF readings for an emitter along the center line. This is because the area under the normal distribution curve out to the standard deviation point

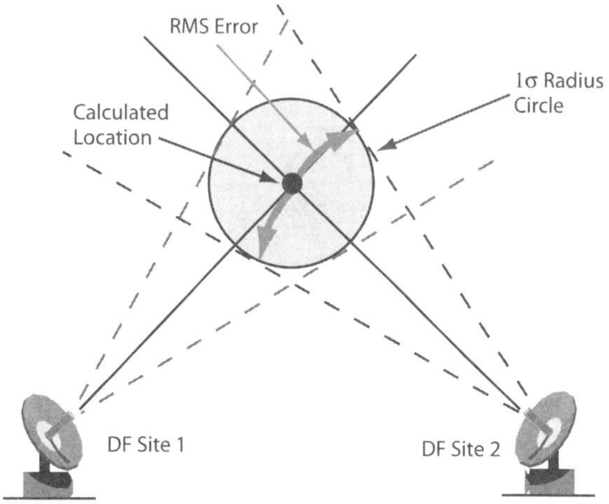

Figure 6.18 With a 90° aspect angle, there is a 46.5% probability that the emitter lies within the 1σ circle centered on the calculated emitter location.

is 0.341. Thus, a circle that just fits inside the RMS lines in this geometry should have 46.5% chance of containing the actual emitter location (0.682 0.682 = 0.465). However, if the radius of this circle is increased to 1.037 of the original circle, it would have a 50% chance of containing the emitter. This is because the area under the normal distribution curve out to 1.037 standard deviation is 35.36% as shown in Figure 6.19 ($[0.3536 \times 2]^2 = 0.5$).

To hopefully clarify the concept of the previous paragraph, consider that there is a 70.7% probability that the actual emitter location is within the $\pm\ 1.037\sigma$ error limits as observed from each direction-finding station. The probability that it is within the same limits for two optimally oriented stations is the square of 0.707 or 0.5.

6.4.3 Elliptical Error Probable

Now consider the less desirable emitter location geometry in Figure 6.20. An ellipse with dimensions 103.6% of the ellipse that just fits inside the RMS error lines would define the EEP. Like CEP, the EEP can be defined for 50% or 90% probability.

6.5 Emitter Location—Error Budget

The most important measure of value of an emitter location system is normally considered to be location accuracy. In the specification of a system, it is necessary to account for all of the elements which contribute error. This is called the

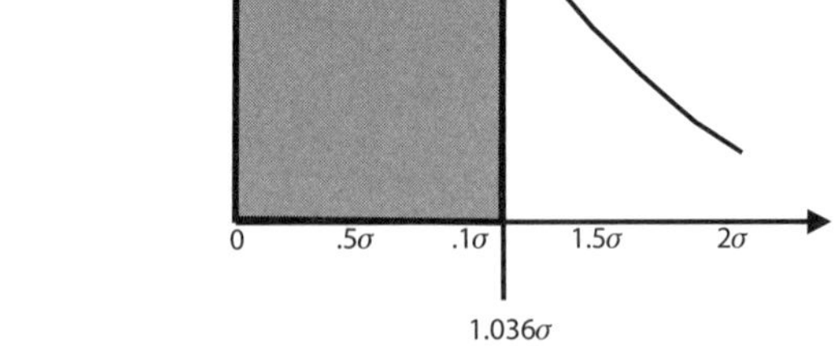

Figure 6.19 For normally distributed errors with zero mean, the area under the Gaussian curve defines the probability that the actual location is within a given radius.

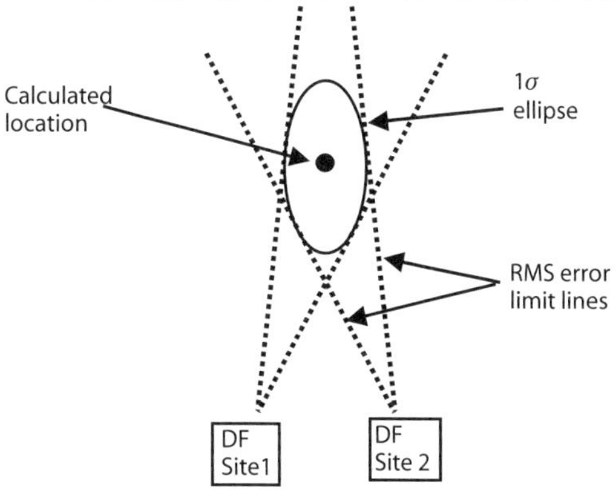

Figure 6.20 There is a 46.5% probability that the emitter lies within the 1σ ellipse in this non-ideal geometry case.

error budget. Some elements are common to more than one type of emitter location approach, but many are associated with only one approach.

6.5.1 Combination of Error Elements

There are multiple sources of emitter location error; some are random and some are fixed. In general, if sources of error are random and independent of each other, they are statistically combined. The total error is the square root of the sum of the squares of the components, as shown here:

$$\text{Total RMS error} = \text{SQRT}\left[\text{error}_1^2 + \text{error}_2^2 + \text{error}_3^2 + \text{error}_4^2 + \ldots + \text{error}_n^2\right]$$

where there are n independent and random sources of error.

However, if the sources of error are not random, they must be summed directly.

As described earlier, the RMS error is a combination of the mean error and the standard deviation. Where a very accurate and complete measurement of system error can be made, for example, on an instrumented range, it may be practical to offset all location measurements or direction-of-arrival measurements by the value of the statistical mean error. Then, the RMS error of the system will be equal to the standard deviation of the measured error data. Note that this assumes no significant site errors—but rather that the major sources of error are associated with the platform. Direction-of-arrival systems on airborne platforms will usually have this characteristic, since reflections from the airframe can cause significant angle-of-arrival errors, and more distant multipath reflectors cause less measurement error.

6.5.2 Impact of Reflections on AOA Error

Reflectors near the path from the target emitter to the AOA site cause errors by creating multipath. The AOA site measures the vector sum of the direct path component and all multipath components arriving at its antennas. As shown in Figure 6.21, reflectors near the target emitter cause multipath signals to arrive at relatively small offset angles which cause relatively small errors (typical in airborne systems). However, reflectors near the AOA site cause multipath signals which can arrive at relatively large angles. These reflections cause relatively large errors. Ground-based AOA systems are significantly impacted by close-in terrain. However all AOA systems have significant multipath errors from the vehicles (air or ground) on which they are mounted. Reflections from the sides (versus close or opposite to the signal AOA) cause the most significant errors.

6.5.3 Station Location Accuracy

For any kind of emitter location approach, the error in the definition of the location of measuring stations must be directly added to the emitter location error as shown in Figure 6.22. If airborne, shipboard or ground mobile measurement systems are used, the station location is taken from the inertial navigation systems on the platforms. Until a few years ago, aircraft INS location accuracy degraded as a function of time since the aircraft left a fixed location on an airfield or on a carrier deck. However, modern INS units use GPS reference for continuous location calibration. This has greatly improved the accuracy over long missions. Since ships have excellent navigation capability, the location error for shipborne measurement sites is extremely small.

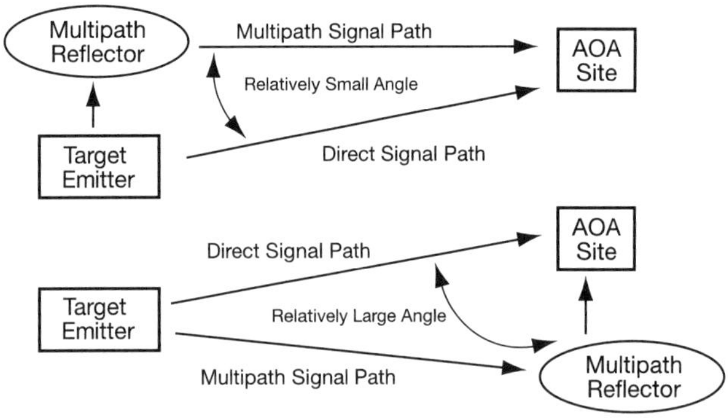

Figure 6.21 Reflectors close to the target emitter cause small AOA measurement errors, while reflectors close to the AOA system cause large AOA errors.

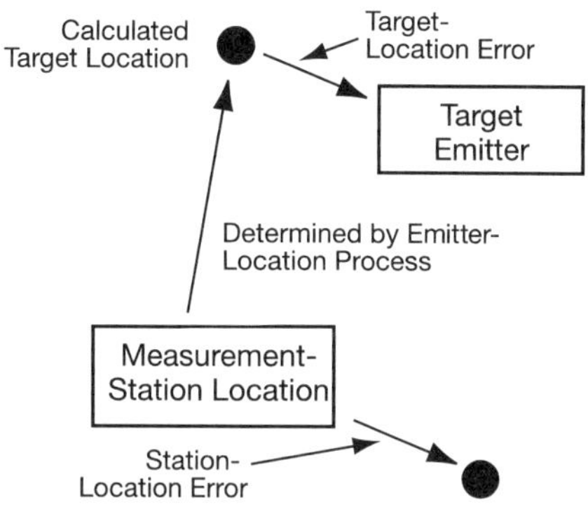

Figure 6.22 Errors in measurement station location transfer directly to errors in target-emitter location.

Fixed ground sites are in locations surveyed to very tight accuracy. The location error is basically only from static or dynamic flexing of towers. Ground mobile measurement sites had relatively crude location until the availability of GPS, since INS for "low-value platforms" was impractical. It was necessary to stop and set up in known map locations to achieve accurate site location. However, GPS receivers are small and provide excellent location accuracy. Thus, even measurement sites in trucks or carried by dismounted individuals can be located to a very few meters.

6.5.4 Error Budget Items for Angle-of-Arrival Emitter Location Approaches

AOA systems measure the angle (azimuth and/or elevation) between the direction from which the signal arrives at a measurement site and some reference angle. The error-budget components include the accuracy of the angle measurement and the accuracy of the directional reference as shown in Figure 6.23. Since elevation is relative to the local vertical or horizontal, the reference is easy to acquire to very high accuracy. The azimuth reference (typically true North) presents more of a challenge. In high-value platforms (e.g., ships and aircraft) the angular reference comes from the INS. The INS has directional gyroscopes which drift, but modern systems calculate the azimuth vectors between subsequent GPS location fixes to update the angular reference for long-term accuracy. There are now small, light INS units for low-value platforms. They use fiber-optic gyroscopes for angular reference and GPS for location. While not as accurate as the larger INS units, they provide high-quality directional reference and location.

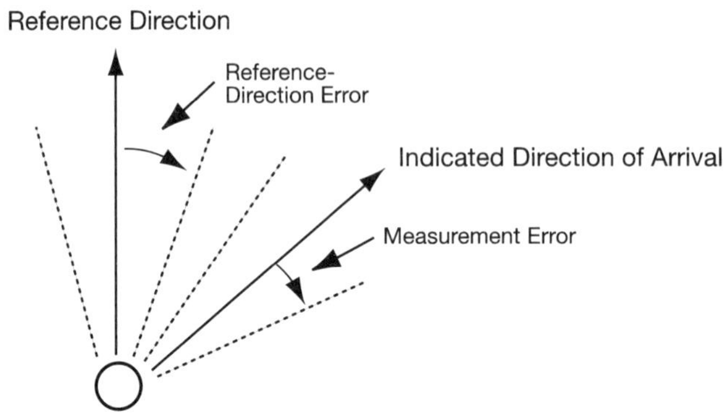

Figure 6.23 When one of the direction-of-arrival techniques is used, the actual direction of the arrival of the signal will be the sum of the measurement of error and the error in the reference direction (e.g., true North).

In earlier ground mobile platforms, it was necessary to use a magnetometer for the north reference. This device senses the Earth's magnetic field. It can be thought of as the equivalent of a magnetic compass that can be read electronically. The magnetometer was mounted in the measurement system antenna array and thus measured the actual orientation of the array, even if it moved with the wind or with guy wire tension. However, the magnetometer suffered from the same problems as a manual compass (e.g., declination variable with location) and had a relatively low accuracy (about 1.5° RMS).

Another source of error in AOA systems is the orientation of the antenna array relative to the angular reference. Unless the angular reference device is mounted on the array, there will be an offset error each time the array is deployed (if it is on a mast) or when it is mounted on an airborne platform.

6.5.5 Error Related to Signal-to-Noise Ratio

It is typical to specify the location accuracy for a system receiving a strong signal, however the system must typically handle much weaker signals. One way to specify the sensitivity of a direction-finding system is to take a series of measurements (usually 5 to 10) at increments of received signal strength. For strong signals, all AOA measurements will be very close (typically identical). Then as the signal strength decreases, the reduced SNR will cause variations in the measured AOA. The system sensitivity is often stated as the received signal strength at which the standard deviation of these measurements equals 1°. It is also possible to calculate the RMS angular error component caused by any specified SNR, but this varies with the specific system configuration.

6.5.6 Calibration Errors

All AOA systems that achieve high accuracy are calibrated to remove the effects of fixed errors from antenna-mounting geometry, vehicle reflections, and processing. The calibration involves measuring AOA on some sort of accurate range and the correction of measured data in later operations to remove the errors measured during calibration. The accuracy of the calibration data is an additional source of angular error.

6.5.7 Combination of AOA System Errors

It is usually reasonable to consider all of the earlier stated error factors except site location as independent and random. Thus, they can be statistically combined to determine the overall error budget. However, there are cases in which some errors must be considered additive. For example, if there is a mean error that is not corrected, it should be considered additive. In some systems, there are also calculation errors which are a fixed function of angle of arrival. These errors are not random, and if not corrected during processing, should be considered additive (versus angle).

6.6 Conversion of AOA Errors to Location Errors

The most important issue concerning the application of an emitter location system is the accuracy with which it can locate an emitter. For any kind of location system, this depends both on the measurement accuracy and the engagement geometry. For the moment, we consider only AOA systems.

6.6.1 Measurement Accuracy

The standard deviation is the angle of the statistical standard deviation from the mean error value. Statistically, that means 34% of the erroneous angle readings are less than this far from the mean error value. However, since the errors can be either to the right or left of the mean, the two standard deviation angles include 68% of the measurement error data.

Since the mean error is heavily influenced by reflection from the vehicle (air, ground, or sea) that is carrying the AOA system, system calibration can be expected to remove most of its impact on the final system RMS error. Thus in calibrated AOA systems, you expect the RMS error to be primarily the standard deviation of the corrected data.

As shown in Figure 6.24, a single AOA site locates the emitter only to a "pie slice" area. If the two RMS lines are taken to be the standard deviation lines, there is considered to be a 68% chance that the emitter is within this angular area. If the RMS error values are low, the wedge is narrow and the angular accuracy is greater. However, the linear error is also a function of distance from the AOA site. The width of the error area in the vicinity of the located emitter can be calculated from:

$$W = 2D\tan(\theta)$$

where

W = the distance from the true angle vector to the RMS error vector (in any units);

D = the distance from the site to the located emitter (same units);

θ = the RMS angular error.

With two AOA sites, the location of the emitter is determined by triangulation. Figure 6.25 shows the intersection of the angular areas from the two sites. Ideally, the two sites would be located 90° apart as viewed from the located emitter. This would minimize the area of location uncertainty. The figure shows a more general case in which the two sites are not ideally located. There is now a

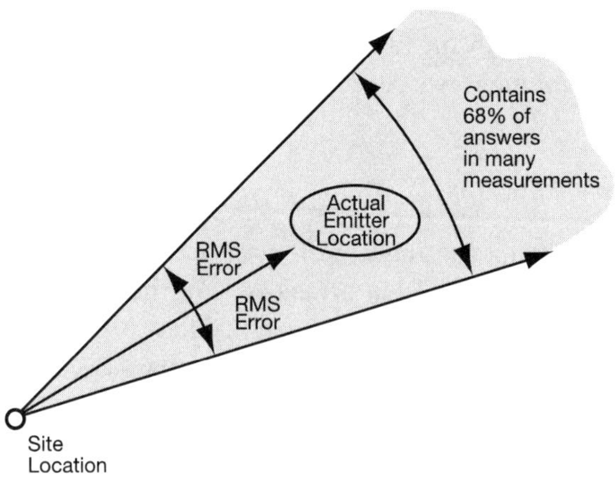

Figure 6.24 The angular area between the two rms error lines contains 68% of the solutions that would be found from many calculations with the normally distributed errors with standard deviation equal to RMS error with mean error removed.

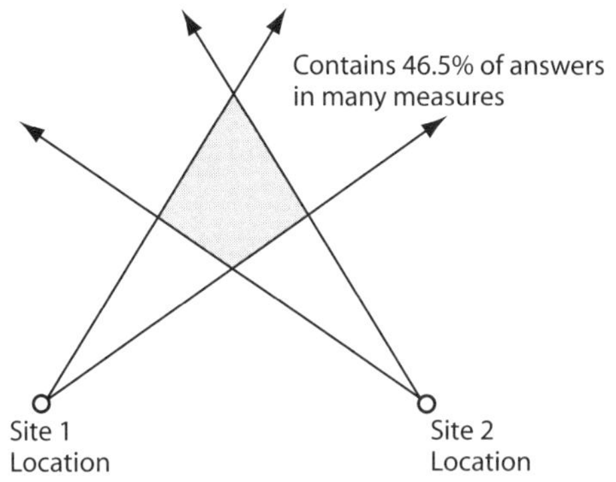

Figure 6.25 The kite-shaped area between both sets of RMS error lines has a 68% chance of containing the emitter location.

68% chance that the located emitter is within the kite-shaped area of the intersection of the two pie slices.

If you were to run a computer simulation of the location of this emitter from these two sites many times, randomly selecting error values, the location plots would form into an elliptical area with high-location density in the center and statistically decreasing density as you move away from the center. The ellipse in Figure 6.26 would contain 46.5% of the plotted solutions.

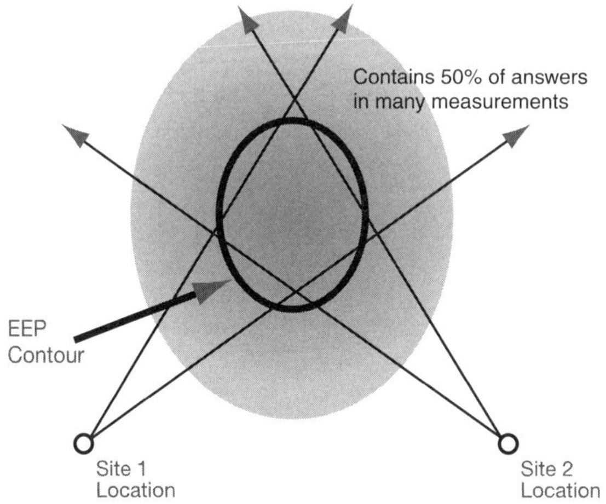

Figure 6.26 The plotted locations from many simulated location measurements with normally distributed error values form an elliptically shaped scatter pattern with 50% of the values within the EEP ellipse.

6.6.2 Circular Error Probable

CEP is defined in Section 6.4.2. For an emitter location system (i.e., multiple AOA sensors and the necessary triangulation processing), we can determine the CEP from the ellipse of Figure 6.26.

First, we need to resize the ellipse from 46.5% inclusion to 50% inclusion. For this, we increase both the major and minor axes of the ellipse by multiplying each by the factor 1.036. This makes an ellipse 50% likely to contain the emitter. Note that this ellipse can be calculated from the specifications of the AOA sites and the intercept geometry. It is often called the "elliptical error probable."

The second step is to calculate the magnitude of the vector sum of the semimajor and semiminor axes of this ellipse (as shown in Figure 6.27) from the formula:

$$CEP = 0.75\left[\sqrt{a^2 + b^2}\right]$$

where

CEP = the radius of the circular error probable circle (any units);

a = the semimajor axis of the ellipse (same units);

b = the semiminor axis of the ellipse (same units).

This value for CEP (said to estimate the true CEP to within 10%) comes from a June 1971 Rand report (R-722-PR) by L. H. Wegner entitled, "On the

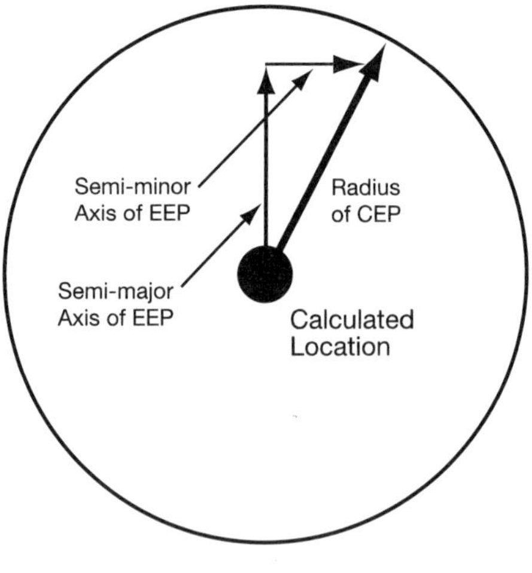

Figure 6.27 CEP is the radius of a circle centered at a measured emitter location that has a 50% probability of containing the actual emitter.

Accuracy Analysis of Airborne Techniques for Passively Locating Electromagnetic Emitters," which is widely used as a reference.

As used in emitter location systems, the calculated elliptical error probable is drawn on a map display around the measured location of an emitter. It is then assumed that there is a 50% probability that the emitter is within the ellipse. This information is extremely valuable to a military analyst who can add other tactical situation and recent history information to determine the tactical situation to appropriate accuracy.

CEP, on the other hand, is most useful in comparatively evaluating different systems and tactics in realistic geometric scenarios.

6.7 Location Errors in Precision Location Systems

TDOA and FDOA are the precision emitter location approaches discussed in Section 6.3.

The accuracy with which TDOA and FDOA systems calculate emitter locations is usually stated in terms of the CEP (as defined earlier), however the CEP is determined in terms of the accuracy with which the isochrones and isofreqs are calculated. Like location accuracy discussed for AOA systems, the CEP for TDOA and FDOA depends on the accuracy of measurements and geometry. The accuracy calculations are based on the assumption that many independent measurements are taken. Thus, the accuracy is statistically defined.

6.7.1 TDOA System Accuracy

Figure 6.28 shows the locations of an emitter and two TDOA receiving stations. The isochron is the curve that includes all of the possible locations in the plane that would cause the measured time difference of arrival (with the signal traveling at the speed of light). The location-error budget includes the accuracy of the locations of the sites and the accuracy with which the time is measured. The following equations calculate the standard deviation of the accuracy with which this hyperbola is drawn, assuming that the site locations are exact. Thus, it is only dependent on the intercept geometry and the time measurement accuracy. The time measurement error is assumed to be Gaussian. This is sometimes described as the "thickness" of the hyperbolic line. Rather than put this into one huge equation, we will do it in a couple of steps.

As shown in Figure 6.28, there are two sites, separated by a distance B_1 along the x-axis of the coordinate system. The emitter is at location X, Y in kilometers relative to the coordinate origin located at site 1.

First, we need to calculate the lengths of the two signal paths:

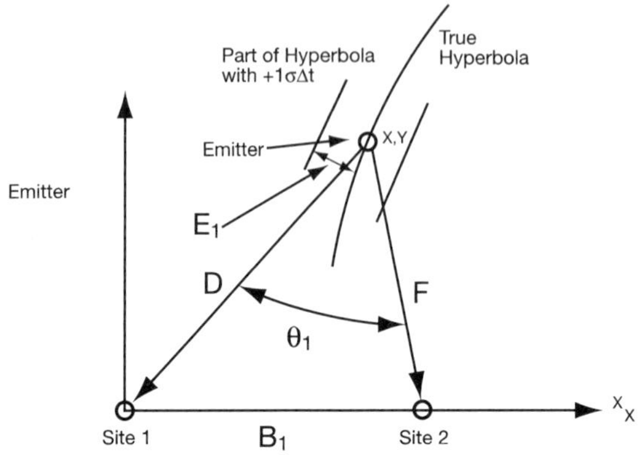

Figure 6.28 The TDOA for signals arriving at sites 1 and 2 define a hyperbola passing through the emitter. A 1σ TDOA error causes the hyperbola to move a distance E_1.

$$D = \text{sqrt}\left(X^2 + Y^2\right)$$

$$F = \text{sqrt}\left(\left[B_1 - X\right]^2 + Y^2\right)$$

Then we calculate half of the sum of the sides of this triangle (S_1):

$$S_1 = \left(D + F + B_1\right)/2$$

Next, we calculate the sine of the half angle between sites 1 and 2 (θ_1) as seen from the emitter.

$$\sin\left(\theta_1 / 2\right) = \text{sqrt}\left(\left[S_1 - D\right] * \left[S_1 - F\right] / \left[D * F\right]\right)$$

Now we can write the equation for the standard deviation (1σ) error of the offset (in kilometers) of the erroneous hyperbola from the true hyperbola that passes through the emitter. (The offset distance at the signal location is called E_1.) The Δt term is the 1σ error of the TDOA measurements.

$$E_1 = 0.00015 * \Delta t / \sin\left(\theta_1 / 2\right)$$

6.7.1.1 TDOA CEP

As described in Section 6.3.1, the hyperbola is just a line through the emitter, a second hyperbola must be drawn from another baseline to find the location at the intersection. Figure 6.29 shows the engagement with a third intercept site. We will calculate the hyperbola offset from a second baseline and draw the resulting error area. For simplicity, we will place the third site along the *x*-axis, a distance B_2 beyond site 2. Again, the calculations are made in steps, beginning with the calculation of the third signal path length.

$$G = \text{sqrt}\left(\left[B_1 + B_2 - X\right]^2 + Y^2\right)$$

Then calculate half of the sum of the sides of the second triangle (S_2):

$$S_2 = \left(F + G + B_2\right) / 2$$

Next, we calculate the sine of the half angle between sites 2 and 3 (θ_2) as seen from the emitter.

$$\sin\left(\theta_2 / 2\right) = \text{sqrt}\left(\left[S_2 - D\right] * \left[S_2 - F\right] / D * F\right)$$

Now we can write the equation for the standard deviation (1σ) error of the offset (in kilometers) of the erroneous hyperbola from the true hyperbola that passes through the emitter. (The offset distance at the signal location is called E_1.) The Δt term is the 1σ error of the TDOA measurements.

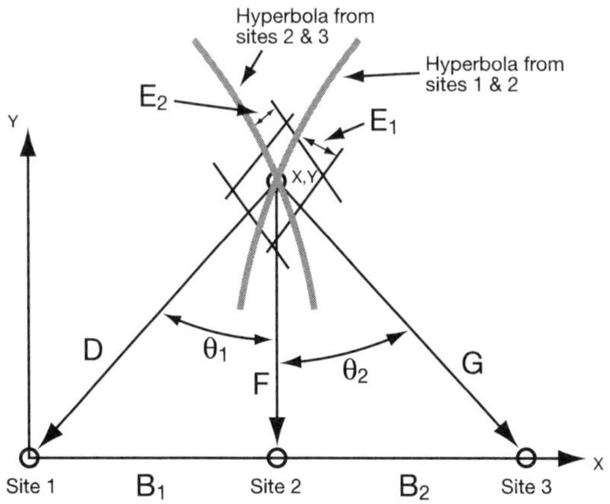

Figure 6.29 With a third site, a second hyperbola is defined. The two hyperbolas intersect at the emitter. A set of $\pm 1\sigma$ hyperbola offsets can be calculated from the baseline B_1.

$$E_2 = 0.00015^*\Delta t / \sin(\theta_2 / 2)$$

The area enclosed by the intersections of the $\pm 1\sigma$ lines about each hyperbola at the emitter location is the parallelogram shown in Figure 6.30. As in the case of the AOA CEP calculation, a computer simulation run with random TDOA measurement errors (Gaussian distribution) would generate the elliptical point density scatter plot as shown in Figure 6.31. The ellipse containing 50% of the solutions is the EEP, and the CEP can be calculated from the EEP as described in Section 6.6.2.

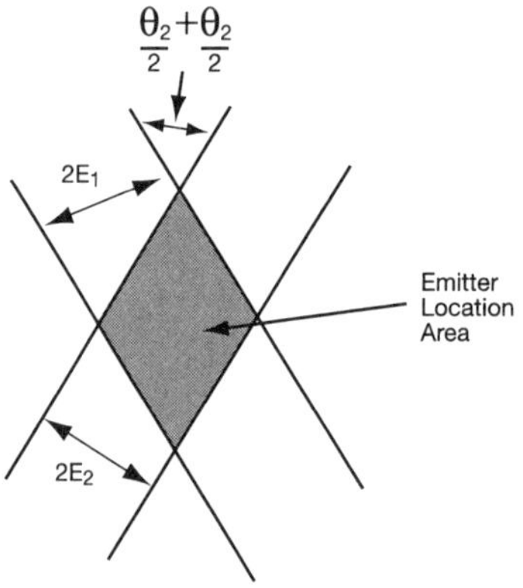

Figure 6.30 The $\pm 1\sigma$ error lines from the two TDOA baselines cross at an angle that is the sum of the two half-angles to the sites as observed from the emitter, forming a parallelogram.

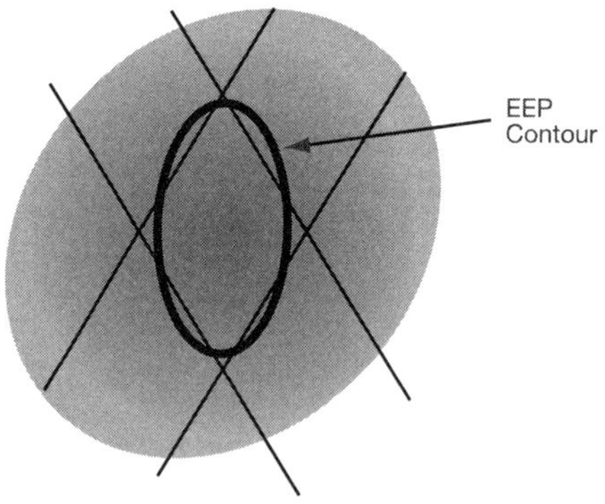

Figure 6.31 The plotted locations from many simulated TDOA measurements with normally distributed error values form an elliptical pattern. The ellipse containing 50% of these solutions is the EEP.

6.7.2 Location Errors in FDOA Emitter Location Systems

Like the TDOA systems discussed earlier, FDOA system accuracy is considered in terms of the location CEP or EEP. Like AOA and TDOA, the accuracy calculations are based on the assumption that many independent measurements are taken.

The formula for isofreq accuracy presented here is based on a classic article by Dr. Paul Chestnut in the March 1982 issue of *IEEE Transactions on Aerospace and Electronic Systems.* The formulas given here assume that the emitter is far enough from the sensor platforms that the depression angle to the emitter from the receiving platforms can be ignored. It also assumes that the location and velocity vector of each receiving platform is precisely known. Finally, we have simplified the problem with the assumption that the receiver platforms are flying at the same speed in line along a level baseline path. (Please note that Dr. Chestnut's formulas are more complex because they do not make these assumptions.)

6.7.2.1 Isofreq Accuracy

Figure 6.32 shows the locations of an emitter and two FDOA receiving stations. The isofreq is the curve that includes all of the possible locations in the plane that would cause the measured frequency difference of arrival with this intercept situation—as described in Section 6.3.2. The following equations calculate the standard deviation of the accuracy with which this curve is drawn (with the assumptions discussed earlier). Thus, it is only dependent on the intercept geometry and the frequency measurement accuracy. The frequency measurement error is assumed to be Gaussian. This is sometimes described as the "thickness" of the isofreq line. Note that the isofreq curve is perpendicular to the bisector of the angle θ_1 at the true emitter location, so the measurement error causes the line to move in the direction of the bisector. Rather than put this into one huge equation, we will do it in a couple of steps.

As shown in Figure 6.32, there are two sites, separated by a distance B_1 along the x-axis of the coordinate system. The emitter is at location X, Y in kilometers relative to the coordinate origin located at site 1.

First, we calculate the lengths of the two signal paths with the same formulas used for the TDOA case in Section 6.7.1:

$$D_1 = \text{sqrt}\left(X^2 + Y^2\right)$$

$$D_2 = \text{sqrt}\left([B_1 - X]^2 + Y^2\right)$$

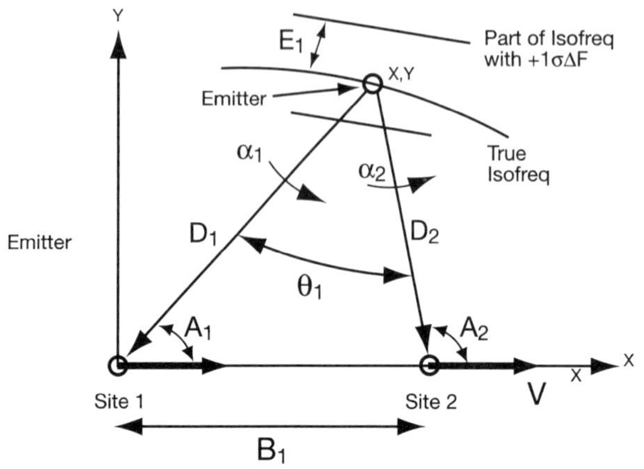

Figure 6.32 The FDOA for signals arriving at sites 1 and 2 define a curve (called the isofreq) passing through the emitter. A $\pm 1\sigma$ FDOA error causes the isofreq to move a distance E_1 in the direction of the bisector of angle θ_1.

Then we calculate the angles between the emitter direction and the velocity vector of each receiving platform):

$$A_1 = \text{arc cos}\left(x / D_1\right)$$

$$A_2 = \text{arc cos}\left(\left[B_2 - x\right] / D_2\right)$$

Next, we calculate the angular velocity ($\hat{\alpha}$) of the line from the emitter to each of the moving receiver platforms:

$$\hat{\alpha}_1 = V \sin\left(A_1\right) / D_1$$

$$\hat{\alpha}_2 = V \sin\left(A_2\right) / D_2$$

Now, we can write the equation for E_1, the standard deviation (1σ) error of the offset (in kilometers) of the erroneous isofreq curve from the true isofreq that passes through the emitter, where F is the transmitted frequency in hertz and ΔF is the 1σ error of the FDOA measurement (in hertz).

$$E_1 = \left(3 \times 10^5 \times \Delta F\right) / \left(F \times \text{sqrt}\left(\hat{\alpha}_2^2 - 2\hat{\alpha}_1\hat{\alpha}_2 - \hat{\alpha}_1^2\right)\right)$$

Note that both the true and erroneous isofreq lines are perpendicular to the bisector of angle θ_1.

$$\theta_1 = A_2 - A_1$$

6.7.2.2 CEP of TDOA Location

Since the location requires a third measurement site to create a second isofreq which crosses the first at the emitter location, consider Figure 6.33.

We need to calculate D_3, A_3, and $\hat{\alpha}_3$ and θ_2 as earlier:

$$D_3 = \text{sqrt}\left(\left[B_1 + B_2 - X\right]^2 + Y^2\right)$$

$$A_3 = \text{Arc cos}\left(\left[B_1 + B_2 - X\right] / D_3\right)$$

$$\hat{\alpha}_3 = V \sin(A_3) / D_3$$

$$\theta_2 = A_3 - A_2$$

Now we can calculate the error in the second isofreq line from:

$$E_2 = \left(3 \times 10^5 \times \Delta F\right) / \left(F \times \text{sqrt}\left(\hat{\alpha}_3^2 - 2\hat{\alpha}_2\hat{\alpha}_3 - \hat{\alpha}_2^2\right)\right)$$

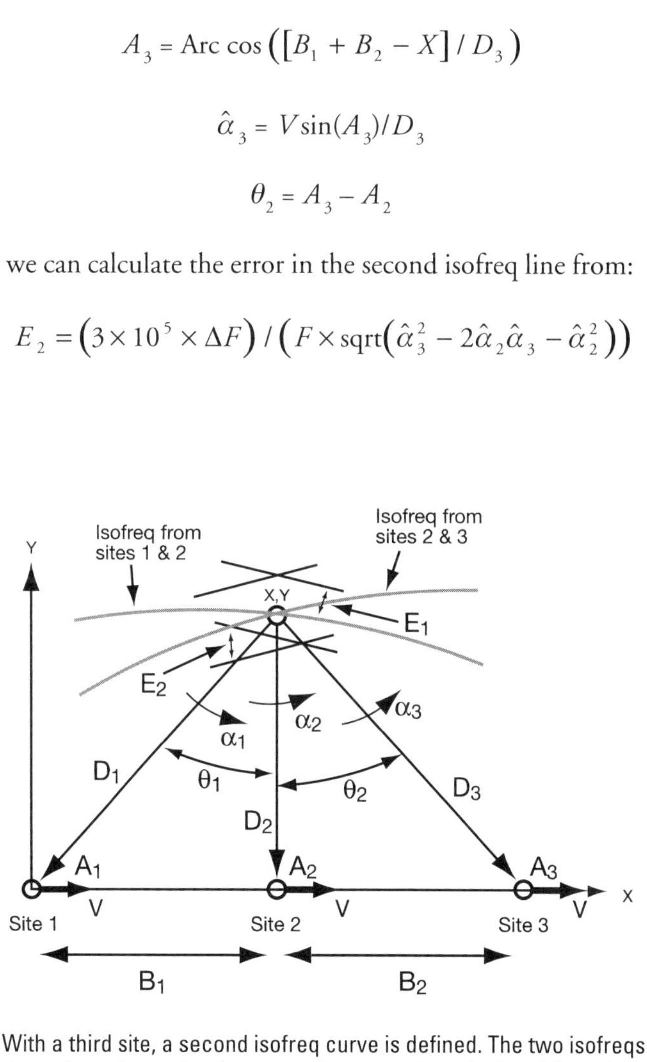

Figure 6.33 With a third site, a second isofreq curve is defined. The two isofreqs intersect at the emitter. A set of $\pm$ 1σ curve offsets can be calculated from the baseline, B_2.

The parallelogram of Figure 6.34 shows the $\pm 1\sigma$ erroneous isofreq lines from the two FDOA baselines. Since the errors (E_1 and E_2) are Gaussian, there is a 46.5% probability that the measured value will lie between the $\pm\ 1\sigma$ error lines. If the values of E_1 and E_2 are multiplied by 1.036, the resulting parallelogram would contain half of the data points. As for AOA and TDOA systems, a computer simulation run with Gaussian distributed FDOA measurements would generate the elliptical point density scatter plot as shown in Figure 6.35. The ellipse containing 50% of the solutions and the EEP and the CEP can be calculated as for the AOA case (in Section 6.4).

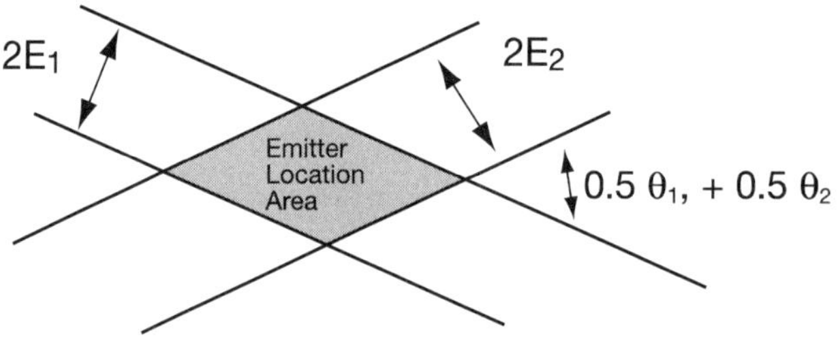

Figure 6.34 The $\pm 1\sigma$ error lines from the two FDOA baselines cross at an angle that is the sum of the two half-angles to the sites as observed from the emitter, forming a parallelogram.

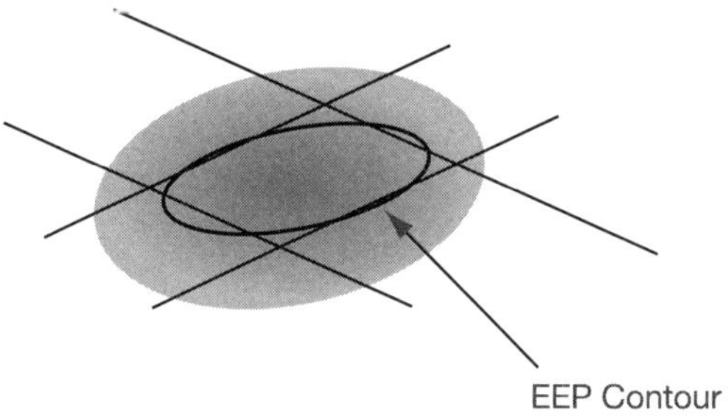

Figure 6.35 The plotted locations from many simulated FDOA measurements with normally distributed error values form an elliptical pattern. The ellipse containing 50% of these solutions is the EEP.

7

Communication Satellite Links

Communication satellites are very much a part of EW. Systems talk to each other over satellite links, and hostile satellite links are logical targets for intercept or jamming. Communication satellites carry information between terminals on or near the surface of the Earth as shown in Figure 7.1. Link equations are defined for the uplink, the downlink, and the whole path between the terminals. Although satellite links use the same laws of physics as ground-to-ground or air-to-ground communication, they are typically designed and described using different formulas. The differences result from the nature of the space environment and from the way that communication satellites are used.

7.1 The Nature of Communication Satellites

A basic assumption used in the normal EW equations (applied within the atmosphere) is that all of the equipment and the transmission medium is at 290K. This works because the kelvin scale goes down to absolute zero and a temperature change adequate to cause 1 dB of change in the link in either direction is way more than that which causes humans to die. However, in space, very low temperatures (near absolute zero) are common. This requires a different way of looking at receiver sensitivity.

Communication satellite links typically have a great deal of bandwidth to serve multiple, simultaneous users, each of which buys only the amount of bandwidth required. This makes link equations in forms independent of

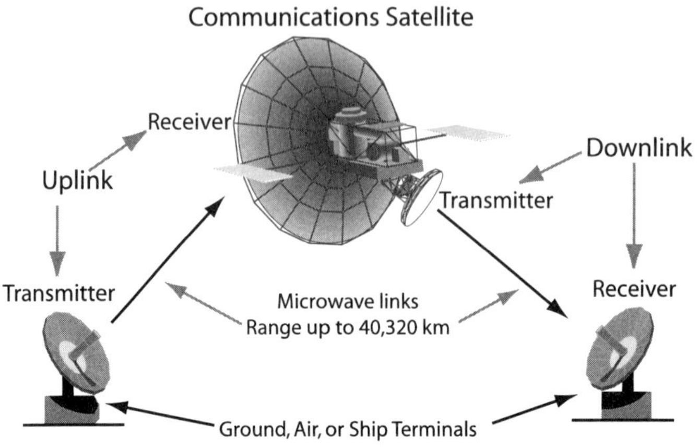

Figure 7.1 Communications satellites carry digital information between points on or near the Earth's surface over microwave links up to 40,320 km in length.

bandwidth very useful. Also, communication satellites carry their information in digital form, so equations often use digital communication terms.

The rest of the differences are just a matter of the way SATCOM type folks are used to communicating.

In this chapter, we will:

- Review the associated terms and definitions;
- Calculate a receiver's noise temperature;
- Discuss satellite orbits;
- Describe the link equation forms for uplinks and downlinks;
- Discuss the vulnerability of satellite links to jamming;
- And finally relate the communication satellite link equations to the equivalent terrestrial forms.

7.2 Terms and Definitions

Table 7.1 lists some decibel definitions used for communication satellite links. The reason for the existence of these specialized decibel units will become clear in the discussion of terms and definitions to which they apply. They will be used in the communication satellite forms of the link equation.

Table 7.1
Decibel Definitions for Communication Satellites

Unit	Definition
K	Abbreviation for degrees Kelvin
dBHz	= decibel value of frequency or bandwidth in hertz
dBW/K	= decibel value of power in watts divided by temperature in degrees Kelvin
dBi/K	= decibel value of antenna gain relative to isotropic divided by temperature in degrees Kelvin
dBW/HzK	= decibel value of power in watts divided by product of bandwidth in hertz and temperature in degrees Kelvin
dBW/m^2	= decibel value of signal power density in watts per square meter

Table 7.2 defines several special terms commonly used in communication satellite forms of the link equation. For each, the symbol, the definition, and the applicable decibel form units are given. All of these terms are in decibel (i.e., logarithmic rather than linear) form. Note that it is common to just use the symbol "K" and the term "kelvins" when referring to degrees Kelvin.

Table 7.2
Special Terms Used in Communication Satellite Forms of the Link Equation

Symbol	Definition	Units
C	Carrier power received	dBW
k	Boltzmann's constant	dBW/HzK
C/kT	Carrier to thermal noise with k	dBHz
C/T	Carrier to thermal noise	dBW/K
E_b/N_0	Energy-per-bit to noise-per-unit bandwidth	dB
EIRP	Equivalent isotropic radiated power	dBW
G/T_s	Figure of merit	dBi/K
PFD	Power flux density	dBW/m^2
Q	System quality factor	dB(W/K)
W	Illumination level	dBW/m^2

"C" is used to denote the radio frequency (predetection) signal power arriving at a receiver. This is a little confusing because it stands for "carrier." The actual radio frequency signal contains a carrier (at the nominal transmission frequency) and modulation sidebands (which carry information). In this case, "carrier" does not really mean carrier, it means the whole signal—carrier and sidebands.

"k" stands for Boltzmann's constant (1.38×10^{-23} watt seconds/K). The lower case k is used for this constant to avoid confusion with the upper case K used for degrees Kelvin. Since the actual units of hertz are 1/second, Boltzmann's constant can also be stated in the linear form units watt/(HzK) or in the decibel units dBW/HzK. The decibel form of Boltzmann's constant (a much-used number) is:

$$-228.6 \text{ dBW/HzK}$$

"C/kT" is an expression for the received carrier to noise power per Hz of bandwidth. Think of kT as kTB (used in EW receiver sensitivity calculations) with the bandwidth term missing. The linear units of this expression would be watts/[(watt sec/K)(K)] which simplifies to 1/second or hertz. Thus, the units of dBHz for C/kT in decibel form.

"C/T" is the ratio of the received signal power to the thermal noise temperature in the environment in which that carrier power is measured. If the denominator were changed to kTB this would become the carrier to thermal noise ratio.

"E_b/N_0" is not at all unique to communication satellite calculations, it is widely used in all digital communications as the digital equivalent of SNR. It is most easily thought of as the output SNR from a digital receiver divided by the bit rate—in a 1-Hz bandwidth. N_0 is the thermal noise per hertz of bandwidth (or kT) and E_b is the energy in a single bit (i.e., the signal power multiplied by the bit duration). Since both the numerator and the denominator have the same units (watt second) this is a straight ratio so the units are just decibels.

EIRP stands for equivalent isotropic radiated power. It is the power that would have to come from a transmitter to an isotropic transmit antenna to create the power actually transmitted from the antenna in use. It is just another way to say ERP. Both are signal strength values derived by increasing the transmitter output power by the antenna gain in the direction of the receiver. Its units are dBW.

"G/T_S" is a figure-of-merit term for a receiver in terms most meaningful in communication satellite receiver system design. This is the receiving antenna gain divided by the receiving system noise temperature. It is directly related to the signal level that must arrive at a satellite or Earth station location if a "signal equal to noise" condition is to be achieved—independent of bandwidth. Its units are dBi/K.

"PFD" is a term for the power flux density out in space. A related term is PFD_B which is the power flux density within a specified bandwidth. Since the power flux density which can be radiated from a satellite onto the Earth's surface is defined by the International Radio Consultative Committee (CCIR) in terms of the flux density per 4 kHz of bandwidth, PFD_B is usually stated in units of dBW/m^2 per B_{CCIR} (i.e., per 4-kHz bandwidth).

"Q" is a system quality factor term combining the EIRP with the receiver figure of merit (EIRP + G/T_S). It is related to the received signal quality (independent of bandwidth) without considering propagation losses. It is a convenient way to isolate the losses from other link considerations. Note that the sum of numbers with units of dBW and dB/K will have the units dB(W/K) because the dB term in the numerator of the second term is a unitless ratio.

"W" is the power flux density arriving at the receiving antenna. The units of W are the same as those of PDF (dBW/m^2).

7.3 Noise Temperature

In most EW applications, the transmitters and receivers and all propagation paths are within the atmosphere, so we assume that everything is at (or very near) 290K. This means that we calculate receiver sensitivity in dBm as a function of kTB, the receiver system noise figure, and the required SNR. We calculate kTB from the formula:

$$kTB = -114 \text{ dBm} + 10\log_{10}\left(\text{effective receiver bandwidth} / 1\text{ MHz}\right)$$

As discussed earlier, the satellite link equations do not typically consider receiver sensitivity as such, but do include the receiver system noise temperature at the output of the antenna.

7.3.1 System Noise Temperature

The receiver noise temperature (T_S) is calculated from the formula:

$$T_S = T_{ANT} + T_{LINE} + \left(10^{L/10}\right)T_{RX}$$

where

T_S = the system noise temperature (K);

T_{ANT} = the antenna noise temperature (K);

T_{LINE} = the noise temperature component from the line to the receiver (K);

T_{RX} = the receiver noise temperature (K).

The three component temperatures are determined as shown in the following sections.

7.3.2 Antenna Noise Temperature

The antenna noise temperature is determined by what falls within the antenna beam. If the antenna is pointed at the sun, the noise temperature is so high that the system is, in general, useless until the sun leaves the beam. If the antenna beam is completely occupied by the Earth or by rain, the antenna temperature is approximately 290K. If the full antenna beam is above the horizon, the antenna's side lobes are far below the main-beam gain, and the sky is clear, the antenna temperature is determined from the graph shown in Figure 7.2 (from L. V. Blake's, *Radar Range Performance Analysis*, and earlier NRL reports). Note

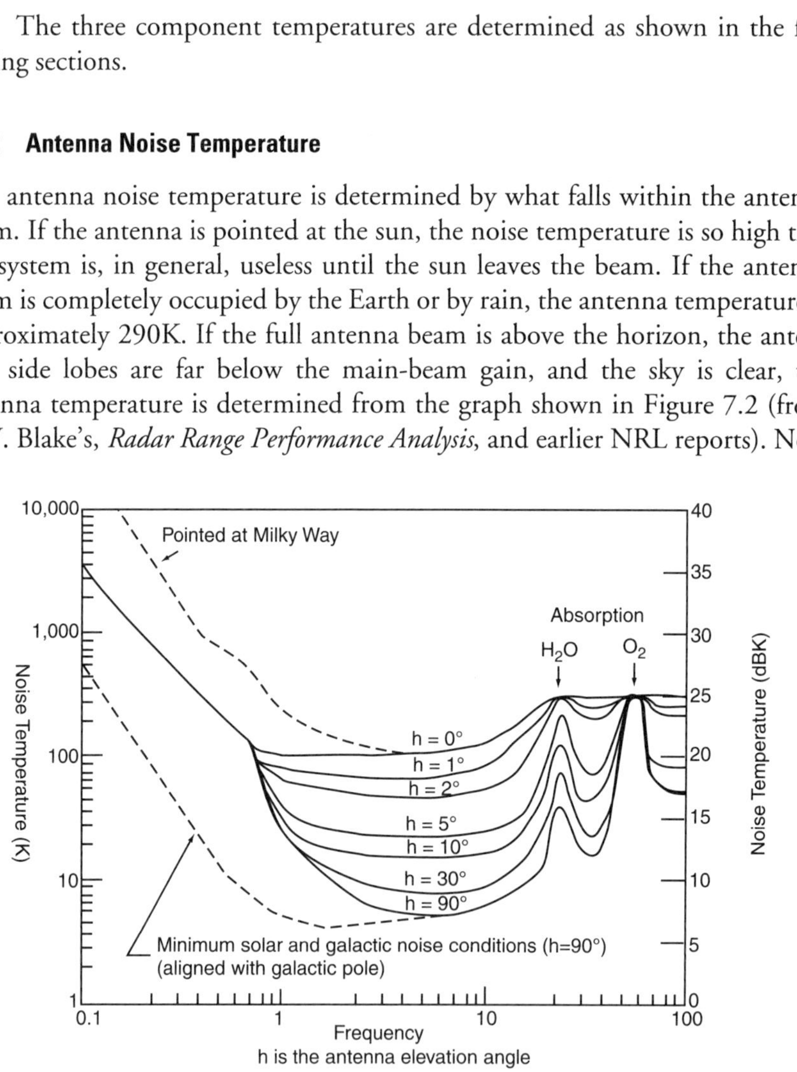

Figure 7.2 The antenna and temperature is a function of frequency and the antenna's elevation angle.

that this graph gives the noise temperature as a function of the receiving antenna elevation and the frequency to which the receiver is tuned. At lower frequencies, the noise emitted by the stars in our galaxy is important, and increases the antenna temperature. The Milky Way is the edge of our galaxy, and thus has the maximum noise temperature.

7.3.3 Line Temperature

The noise temperature of losses between the antenna and the receiver contribute to the system noise temperature according to the formula:

$$T_{LINE} = \left[10^{(L/10)} - 1\right]T_M$$

where

L is the amount of loss before the receiver (in decibels);

T_M is the ambient temperature of the loss mechanism (typically 290K).

Note, that this formula is often shown as:

$$T_{LINE} = \left[T_{ANT} + (L-1)T_M\right] / L$$

where L is the attenuation in linear (versus decibel) form.

Note that the antenna temperature can be significantly lower than the ambient temperature of the attenuator.

7.3.4 Receiver Noise Temperature

The receiver system noise figure can be calculated from its noise figure by the following formula.

$$T_{RX} = T_R\left[10^{(NF/10)} - 1\right]$$

where

T_{RX} = the noise temperature of the receiver (in Kelvin);

T_R = the reference temperature (typically 290K);

NF = the receiver noise figure (in decibels).

If the reference temperature is 290K, the noise temperature can be found from the noise figure by using Table 7.3.

Table 7.3
Noise Figure Versus Noise Temperature

Noise Figure (dB)	Noise Temperature (K)
0	0
0.5	35
1	75
1.5	120
2	170
2.5	226
3	289
3.5	359
4	438
4.5	527
5	627
5.5	739
6	865
6.5	1,005
7	1,163
7.5	1,341
8	1,540
8.5	1,763
9	2,014
9.5	2,295
10	2,610
11	3,361
12	4,306
13	5,496
14	6,994
15	8,881
16	11,255
17	14,244
18	18,008
19	22,746
20	28,710
21	36,219
22	45,672

Where the receiver has multiple gain elements, the receiver noise temperature is dominated by the noise temperature of the first element. The noise temperature component from each downstream stage is reduced by the gain of the upstream stages. In the three-stage receiver shown in Figure 7.3, the receiver noise temperature is found from the formula:

$$T_{RX} = T_1 + (T_2 / G_1) + T_3 / G_1 G_2$$

where

T_1 is the noise temperature of stage 1;
G_1 is the gain (not in decibels) of stage 1;

And so forth.

For each stage, the noise temperature is determined from the T_{RX} formula above.

7.3.5 A Noise Temperature Example

Consider the ground station receiving system shown in Figure 7.4. Since it is in the atmosphere, the reference temperature will be 290K. The receiver is operating at 5 GHz and its antenna is elevated 5°, so we can determine from Figure 7.2 that the antenna temperature is 30K. There is 10-dB loss before the receiver, so the line temperature is 0.9 × 290 = 261K. The receiver noise temperature is 438K from the first stage and 261K from the second stage for 699K total.

The system noise temperature is then 30K + 261K + 699K = 990K.

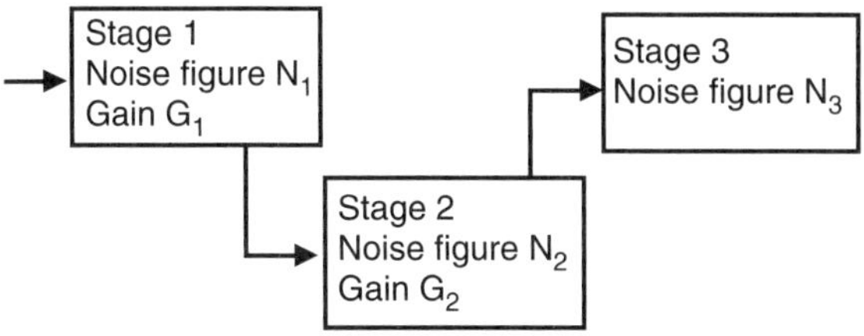

Figure 7.3 The noise temperature of this three-stage receiver is dominated by the noise figure of the first stage.

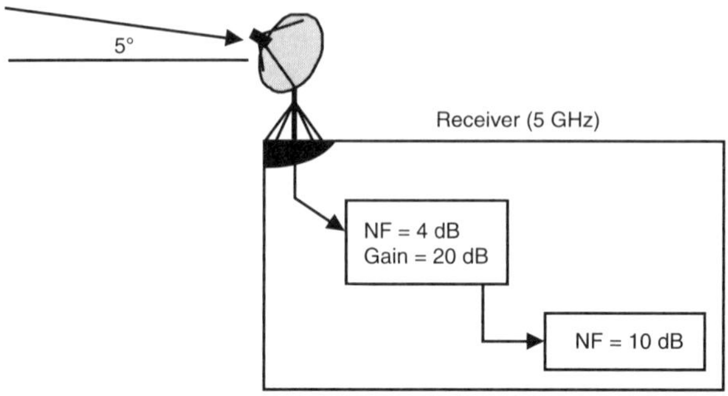

Figure 7.4 This example receiving system has a 5° antenna elevation, 10-dB loss between the antenna and the receiver, and a two-stage receiver tuned to 5 GHz.

7.4 Link Losses

The distances involved in satellite links are quite long, so the link losses are significant. Since much of the link path is outside the Earth's atmosphere, there are also different considerations in the calculation of some of the sources of loss. We will consider: spreading loss, atmospheric loss, and rain or fog loss. Under most circumstances, the sum of these losses (in decibels) can be considered the total link loss. Losses like antenna misalignment are handled separately. Each of these losses applies to the links between the satellite and ground station, and also applies to losses in terrestrial transmitters intercepted by a satellite-based receiver payload.

7.4.1 Spreading Loss

We will calculate the spreading loss in terms of the transfer function between two isotropic (i.e., 0-dB gain) antennas. This is the same formula we have been using for line-of-sight links.

$$L_S = 32 + 20 \log F + 20 \log d$$

where

L_S = the spreading loss in decibels between isotropic antennas;

F = the transmitted frequency in megahertz;

d = the distance in kilometers between the transmitting and receiving antennas.

7.4.2 Atmospheric Loss

Since satellite links pass through the entire atmosphere, we don't consider the loss per kilometer of link as we do in terrestrial links. The loss through the whole atmosphere is shown in Figure 7.5 as a function of frequency and elevation angle. (This figure is from L. V. Blake's *Radar Range-Performance Analysis,* and earlier NRL reports.) Lower elevations have more atmospheric loss because more of their path is within the atmosphere. The curves in this figure include both water vapor and oxygen losses. You can see the water vapor peak at 22 GHz and the oxygen peak at 60 GHz. The extremely high losses near 60 GHz make this an excellent frequency for satellite-to-satellite communication, with no interference from ground-based signals.

For example, at 10 GHz, there is 3-dB atmospheric loss at 0° elevation and 0.5 dB at 5° elevation.

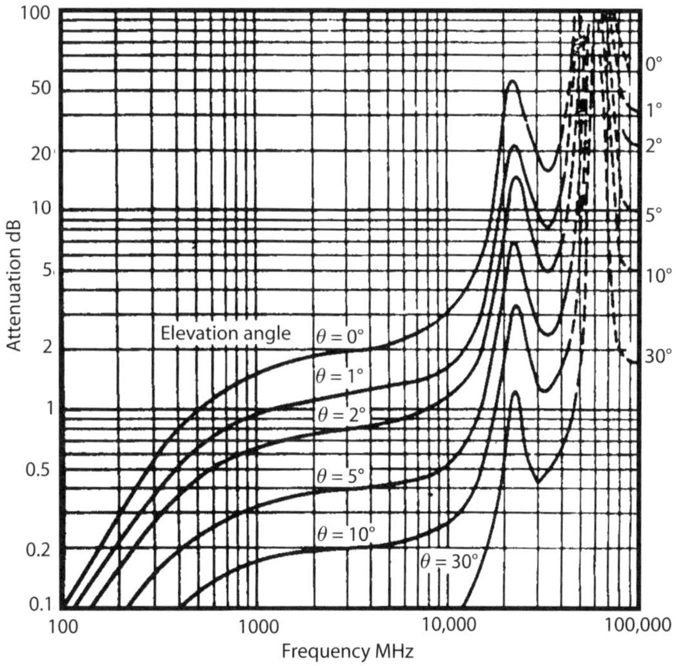

Figure 7.5 Atmospheric attenuation through the whole atmosphere is typically defined in terms of frequency and elevation angle from the ground station.

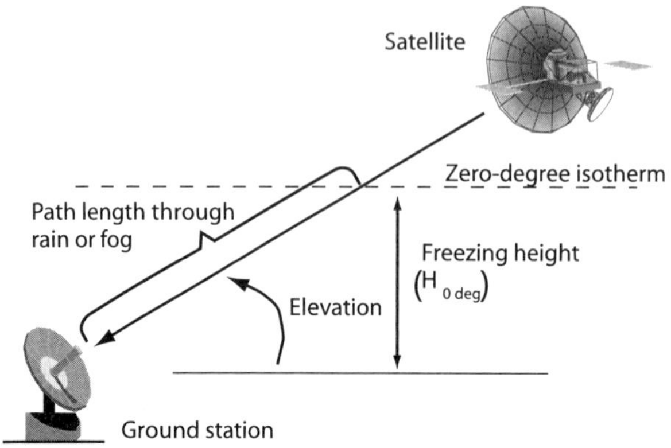

Figure 7.6 The satellite link is subject to rain or fog attenuation from the ground to the 0° isotherm.

7.4.3 Rain and Fog Attenuation

Rain and fog attenuation is more complicated that the other two losses considered. This attenuation is a function of the intensity of the rain or fog, the frequency, and the distance the path is subject to the rain or fog. The geometry of the path-length geometry is shown in Figure 7.6. The transmission passes through the rain or fog from the ground station to the elevation of the 0°C isotherm (at which water starts to freeze). Above this altitude, the precipitation is not water but ice—which causes far lower attenuation. The rain/fog path length is given by the formula:

$$d_R = H_{0\ deg} / \sin El$$

where

d_R = the path length through the rain or fog;
$H_{0\ deg}$ = the height of the 0° isotherm;
El = the elevation angle.

Figure 7.7 is a chart of freezing altitude (0° isotherm) versus latitude. The percentage probabilities indicate the amount of time during the year that the freezing height can be expected to be at or above the indicated altitude.

Once you have determined the path length through the rain/fog, you can determine the amount of rain/fog attenuation from Figure 7.8. This figure is

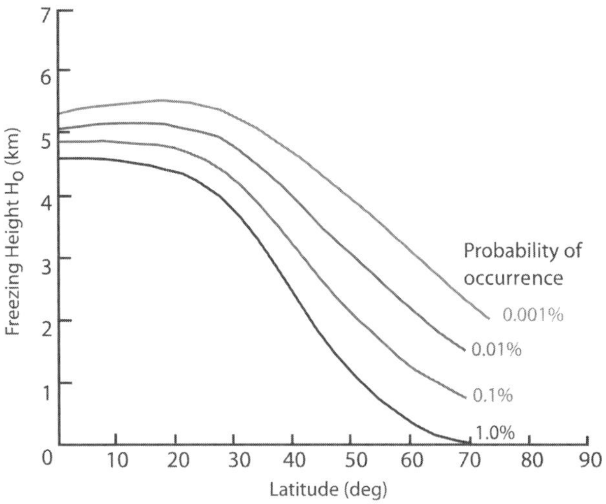

Figure 7.7 The altitude at which water can be expected to freeze is a function of altitude.

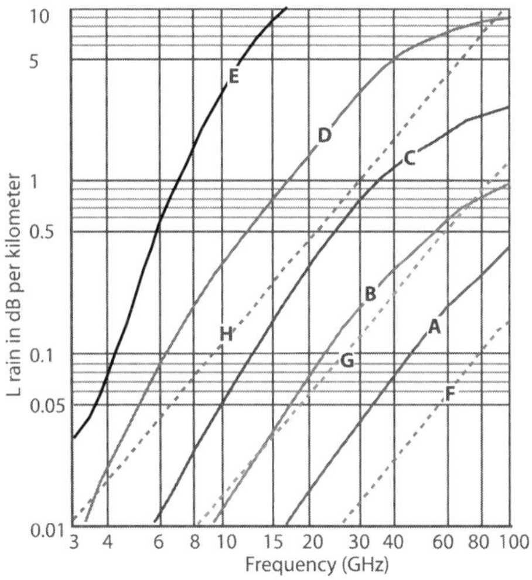

Figure 7.8 Rain or fog attenuation is a function of the density of the rain or fog and the frequency of transmission. Each curve responds to a rain density as defined in Table 7.4 or a fog density defined in Table 7.5.

found in multiple references (including L. N. Ridenour's *Radar Systems Engineering*). This figure should not be interpreted as precise, since the equivalent charts in different references vary by a few decibels in some areas. First select the

Table 7.4
Rain Intensity for Curves in Figure 7.8

A	0.25 mm/hr	Drizzle
B	1.0 mm/hr	Light rain
C	4.0 mm/hr	Moderate rain
D	16 mm/hr	Heavy rain
E	100 mm/hr	Very heavy rain

Table 7.5
Fog Density for Curves in Figure 7.8

F	0.032 gm/m^3	Visibility greater than 600m
G	0.32 gm/m^3	Visibility about 120m
H	2.3 gm/m^3	Visibility about 30m

correct curve for rain from Table 7.4 or for fog from Table 7.5, then determine the attenuation in decibels per kilometer for the operating frequency from that curve (in Figure 7.8). Finally, multiply the path length by the attenuation per kilometer. If there is significant path length through fog, the fog attenuation can be determined by selecting the correct curve from Table 7.5 and applying the appropriate loss for the path length through the fog.

For example, if the latitude is 40° and we are considering the 0.1% condition, the freezing altitude is assumed to be 3 km. If the elevation angle is 30°, the path length through the rain or fog is 6 km (3 km/sin 30). If the link is operating at 10 GHz and there is heavy rain, attenuation would be 2 dB (use curve D; attenuation is 0.33 dB/km × 6 km).

7.4.4 Faraday Rotation

Faraday rotation is caused by the Earth's magnetic field, and causes a rotation in the polarization of signals passing through the ionosphere. This effect is proportional to $1/(\text{frequency})^2$, so the lower the frequency, the greater the polarization loss that can be expected. Very large losses from the Faraday effect can be expected at VHF and UHF, and above about 10 GHz it is normally considered insignificant.

The loss (in decibels) caused by a mismatch between the linear polarization of a received signal and the linear polarization of the receiving antenna is found from the formula:

$$L = -10 \log\{[\text{Cos}(\theta)]^2\}$$

where

L = Loss in decibels;

θ = Polarization mismatch in degrees.

Table 7.6 shows the loss in decibels versus the approximate polarization mismatch. Faraday rotation is variable with time of the day, and other factors which are difficult to predict. However, when matched circularly polarized transmit and receive antennas are used, the polarization loss does not apply.

7.5 Link Losses in Typical Links

Consider two typical satellite communication systems: One using a synchronous satellite and one using a satellite in low Earth orbit. The geometry of these two systems will be the basis for later link throughput calculations.

7.5.1 A Synchronous Satellite

There is a relationship between the average radius of the orbit (the semimajor axis of an ellipse with the center of the Earth at one focus) and the period of the orbit. If the orbit is circular (i.e., an ellipse with zero eccentricity), is in the Earth's equatorial plane, and is at an altitude of 36,000 km, the satellite will orbit the Earth once every 23 hours, 56 minutes, and 4.1 seconds. This allows it to remain over a single point on the Earth's surface, which makes it a "synchronous" satellite. (Remember that the Earth needs to turn slightly beyond 360° in 24 hours for one spot on its surface to face the Sun at noon each day.) A

Table 7.6
Polarization Mismatch Versus Polarization Loss

Loss in decibels	0	1	2	3	4	5	6	7	8	9	10	20
Polarization mismatch (degrees)	0	27	37	45	51	56	60	63	67	69	72	84

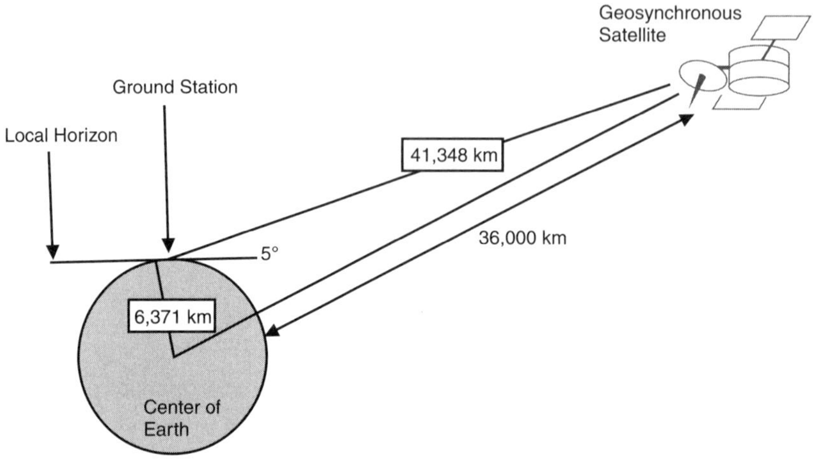

Figure 7.9 A geosynchronous satellite 5° above the horizon from a ground station is about 41,348 km from the station location.

significant advantage of the synchronous satellite is that directional antennas on the Earth need not be steered to maintain pointing toward the satellite.

As shown in Figure 7.9, the distance to a synchronous satellite from a ground station, if the satellite is 5° above the local horizon is 41,348 km. This is

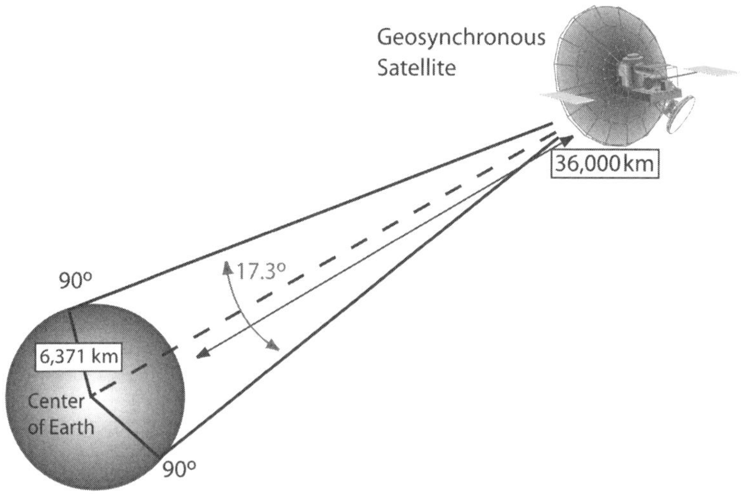

Figure 7.10 An Earth-coverage antenna on a geosynchronous satellite has a beamwidth of 17.3°.

calculated by using the law of sines to determine the sides and angles of the triangle formed by the ground station, the satellite, and the center of the Earth.

7.5.1.1 Earth-Coverage Antenna

It is often convenient to use an "Earth-coverage" antenna as shown in Figure 7.10 on a synchronous satellite. This allows transmission to or from a ground station over all of the Earth that the satellite can see. From the triangle formed by the center of the Earth, the 0° elevation point on the Earth's surface and the satellite, it can be easily proven that the beamwidth from synchronous orbit is 17.3°. If this is the 3-dB beamwidth and the antenna has 55% efficency, the Earth-coverage antenna will have 19.9-dB gain (by the method shown in Section 3.3.4 of *EW 101*).

7.5.1.2 Link Losses to Synchronous Satellite

The link losses to a satellite include spreading loss, atmospheric loss, rain loss, and a few miscellaneous losses that we will discuss later. The spreading loss is found from the formula:

$$L_S = 32 + 20\log(d) + 20\log(f)$$

where

L_S = the spreading loss in decibels;
d = the link distance in kilometers;
f = the frequency in megahertz.

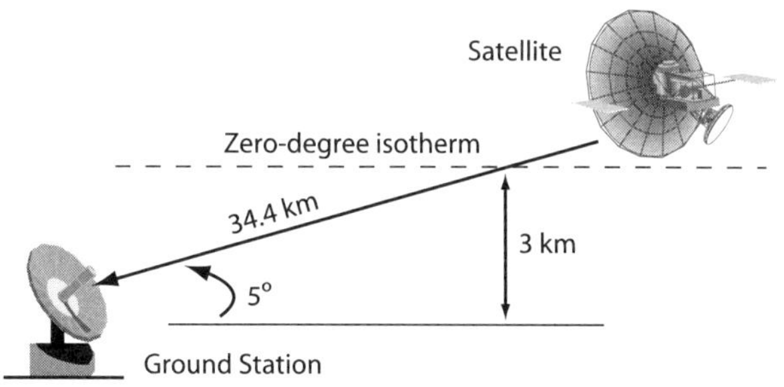

Figure 7.11 For a 5° satellite elevation and a 3-km 0° isotherm, the link is subject to attenuation from moderate rain for 34.4 km.

For a 5° elevated synchronous satellite, the distance is 41,348 km, so at 15 GHz, the spreading loss is 210.2 dB.

The atmospheric and rain attenuation are determined by the methodology presented in Sections 7.3.2 and 7.3.3. The atmospheric attenuation through the whole atmosphere at 5° elevation is 1 dB at 15 GHz as shown in Figure 7.5. If the satellite link is designed to operate in moderate rain from an Earth station at 50° latitude with 0.01% reliability, the rain attenuation is determined from Figures 7.7, 7.8, and 7.11. Height of the 0° isotherm (0.01% reliability) from 50° latitude is shown to be 3 km from Figure 7.7. As shown in Figure 7.11, the slant range to a 3-km elevation at 5° elevation is 34.4 km. Figure 7.8 shows that moderate rain causes 0.15-dB attenuation per kilometer at 15 GHz, so the rain attenuation for a 5° elevated link would be 5.2 dB.

The total link propagation loss is then 216.4 dB.

7.5.2 Low-Earth-Satellite Link

Low Earth satellites have the advantage that they have significantly shorter propagation paths to ground stations. However, at any given time, they can only be seen from a small percentage of the Earth's surface, and Earth stations requiring directional antennas must continuously update antenna pointing to maintain lock on the satellite. To provide coverage (noncontinuous) of a large part of the Earth, low-Earth-satellite orbits are normally inclined relative to the equator. A 90° inclination creates a "polar" orbit which goes over the poles, and thus covers the entire Earth (in several orbits). Since the Earth turns below the orbit of the low-Earth satellite, the Earth trace of each orbit is west of the previous orbit

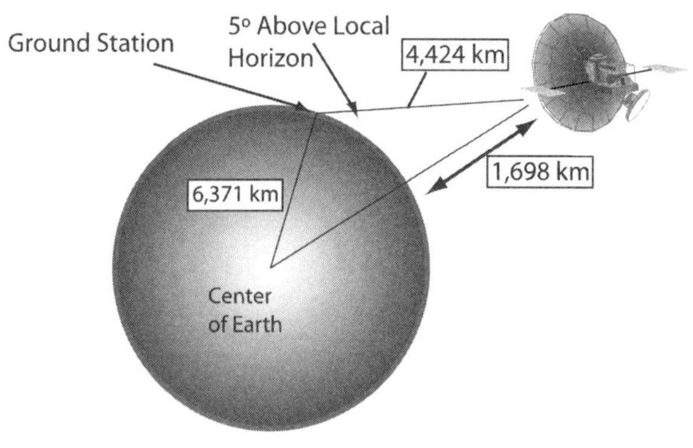

Figure 7.12 A satellite with a 2-hour orbital period that is 5° above the horizon from a ground station is about 4,424 km from the station location.

by 360° (latitude) × the satellite orbital period/23 hours, 56 minutes, and 4.1 seconds.

As shown in Figure 7.12, the distance to a 5° elevated satellite with 1,698-km altitude (i.e., 2-hour orbital period) is 4,424 km. This causes a space loss of 190.8 dB. Because the atmospheric and rain loss occur only in the atmosphere, they will be the same for a 5° elevation angle regardless of the distance to the satellite. Thus, the 1 dB and 5.2 dB determined for the synchronous satellite will also apply to the low-orbit case.

The total link propagation loss is then 197 dB.

7.6 Link Performance Calculations

In this section, we will run the link calculations for the uplink, downlink, and complete throughput for communication from two ground points. We will perform two examples, one for communication via a synchronous satellite and the second via a satellite in a 2-hour Earth orbit. Link distances and losses were calculated for both of these cases in Section 7.5.

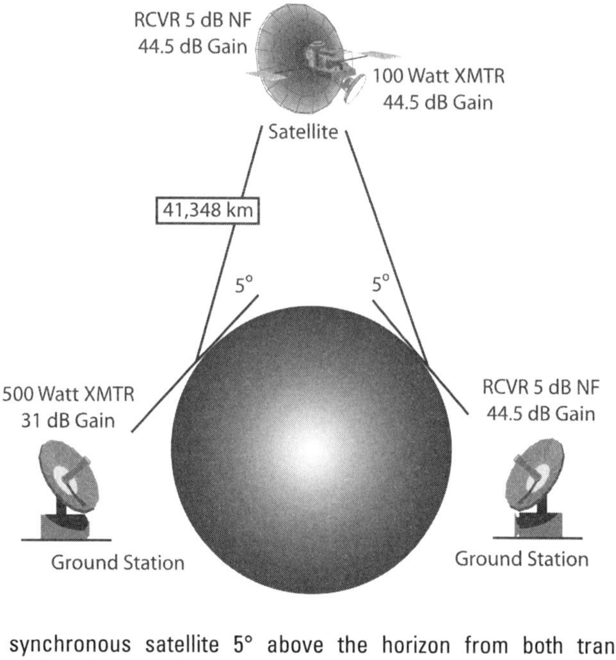

Figure 7.13 A synchronous satellite 5° above the horizon from both transmitting and receiving sites provides point-to-point communication.

7.6.1 Synchronous Satellite Links

As shown in Figure 7.13, the distance from the satellite to each Earth station is 41,348 km, and elevation angles to the satellite from the transmitting and receiving ground stations are both 5°. The transmitter power into the uplink transmit antenna is 500W (+27 dBW), and the antenna gain is 31 dB.

The satellite uplink receiving antenna and downlink transmitting antenna have 44.5-dB gain. The noise figure of the satellite receiving system is 5 dB, including all line losses to the antenna.

The satellite downlink transmitter power is 100W (+20 dBW). We will assume that the operating frequency of both the up and downlinks is 15 GHz. The ground-receiving station has 44.5-dB antenna gain and a receiver noise figure of 5 dB.

As calculated in Section 7.5.1, the link loss for each link is 216.4 dB (210.2-dB spreading loss, 1-dB atmospheric loss, and 5.2 dB allowed for rain loss).

7.6.1.1 Uplink Performance

From Table 7.3, the noise temperature of the satellite uplink receiver is 627K. Since the whole main beam of the uplink-receiving antenna is looking at the Earth, the antenna noise temperature is 290K. This means that the uplink-receiving system noise temperature is the sum of the antenna and receiver noise temperatures, or 917K.

First, let's find the uplink-receiving system figure of merit (G/T_s). The uplink-receiving antenna gain (44.5 dB) in linear form is 28,184. Dividing this by 917K to give G/T_s gives 30.7. Converting this to decibel form gives 14.9 dBi/K.

$$\text{EIRP} = P_T + G_T = +27 \text{ dBW} + 31 \text{dB} = +58 \text{ dBW}$$

The uplink carrier to thermal noise factor is:

$$\begin{aligned} \text{C} / \text{T} &= \text{EIRP} - L + G / T \\ &= +58 \text{ dBW} - 216.4 \text{ dB} + 30.7 \text{ dBi} / \text{K} = -127.7 \text{ dBW} / \text{K} \end{aligned}$$

7.6.1.2 Downlink Performance

Next, let's find the C/T for the downlink. The downlink-receiving system antenna noise temperature, from Figure 7.2 (at 5° elevation for 15 GHz) is 13 dB. The receiver noise temperature (including line effects) is 627 dB. This means that the downlink-receiver noise temperature is the sum of the antenna and receiver noise temperatures, (i.e., 640K).

The downlink receiver figure of merit is then:

$$G / T_S = 28{,}184/640 = 44$$

Converting this to dB form gives 16.4 dBi/K.

The downlink EIRP is:

$$\text{EIRP} = P_T + G_T = +20 \text{ dBW} + 44.5 \text{ dB} = 64.5 \text{ dBW}$$

The downlink losses are the same as the uplink losses (i.e., 216.4 dB).

The downlink carrier to thermal noise is then:

$$\begin{aligned} &C / T = \text{EIRP} - L + G / T_S = +64.5 \,\text{dBW} - 21.6 \,\text{dB} \\ &+16.4 \text{ dBi} / \text{K} = -135.5 \text{ dBW} / \text{K} \end{aligned}$$

7.6.1.3 Combined Uplink and Downlink Performance

The round-trip carrier to thermal noise is found from the formula:

$$1 / (\text{combined}\, C / T) = 1 / (\text{uplink}\, C / T) + 1 / (\text{downlink}\, C / T)$$

This requires converting the C/T factors back to linear form. The uplink C/T is then 1.6982×10^{-13} and the downlink C/T is 2.8184×10^{-14}.

The combined C/T is then $1/(5.8886 \times 10^{13} + 3.5481 \times 10^{13})$ or 1.0597×10^{-15} or –149.8 dBi/K.

To make this meaningful, we need to determine the output SNR that this will provide.

First, we factor in kTB for a 1-Hz bandwidth to find the carrier to thermal noise with k. From Section 7.1, this factor is –228.6 dBW/HzK, so:

$$C/kT = C/T - kTB \text{ for 1-Hz bandwidth} = -149.8 + 228.6 = +78.8 \text{ dBK}$$

Next, we set the bandwidth. For example, let's use 1 MHz.

$$C/N = C/kT - 10 \log B$$

where B is the bandwidth in hertz.

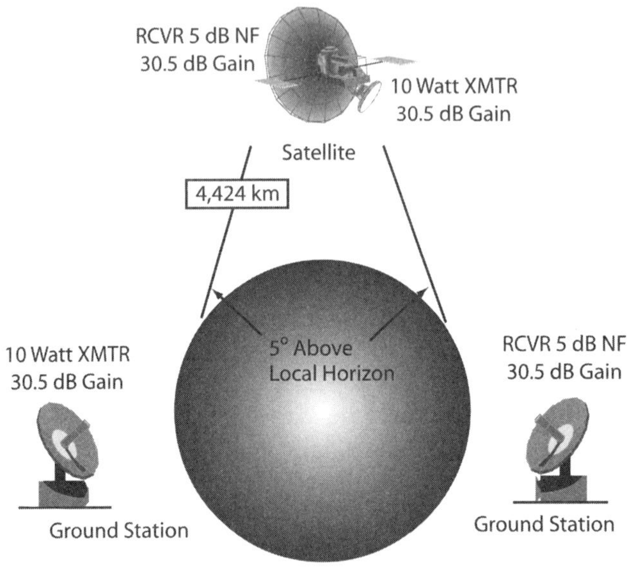

Figure 7.14 A satellite with a 2-hour orbital period that is 5° above the horizon from transmitting and receiving sites is 4,424 km from each station.

$$C/N = 78.8 \text{ dBK} - 10 \log(1{,}000{,}000) = 78.8 - 60 = 18.8 \text{ dB}$$

7.6.2 Low-Earth-Orbit Links

For this example, we'll use the second orbit discussed earlier. The satellite is at 1,698-km elevation. Both the up and downlinks operate at 15 GHz and the transmitting and receiving stations both see the satellite at 5° elevation as shown in Figure 7.14. Both the uplink and downlink antennas on the satellite have 30.5-dB gain, as do the Earth station antennas. The uplink and downlink transmitters each have 10-W output power (+10 dBW). The receivers each have 5-dB noise figures (including line losses to the antennas). As developed in Section 7.5.2, the uplink and downlink losses are each 197 dB.

7.6.2.1 Uplink

The uplink-receiving system figure of merit (G/T_s): the receiver system noise temperature will be 917 dB since the antenna and receiver noise temperatures are the same as for the synchronous satellite case. The uplink-receiving antenna gain (30.5 dB) in linear form is 1,122. Dividing this by 917K to give G/T_s gives 1.224. Converting this to decibel form gives 0.9 dBi/K.

$$\text{EIRP} = P_T + G_T = +10\ \text{dBW} + 30.5\ \text{dB} = +40.5\ \text{dBW}$$

The uplink carrier to thermal noise factor is:

$$C/T = \text{EIRP} - L + G/T_S = +40.5\,\text{dBW}$$
$$-197\ \text{dB} + 0.9\ \text{dBi/K} = -155.6\ \text{dBW/K}$$

7.6.2.2 Downlink

The downlink-receiver system noise temperature is 640K, and the antenna gain is 30.5 dB, just as in the synchronous satellite case, so the receiver figure of merit is also 0.9 dB.

The EIRP is +40.5 dBW, just as for the uplink.

$$C/T = \text{EIRP} - L + G/T_S = +40.5 - 197 + 0.9 = -155.6\ \text{dB}$$

7.6.2.3 Combined Uplink/Downlink Performance

The combined C/T is –152.6 dB (the inverse of the sum of the inverses of the uplink and downlink C/T converted to linear form).

$$C/kT = C/T - kTB \text{ for 1-Hz bandwidth} = -152.6 + 228.6 = +76\ \text{dBK}$$

$$C/N = C/kT - 10 \log B$$

In a 100-kHz bandwidth, this is 16 dB.

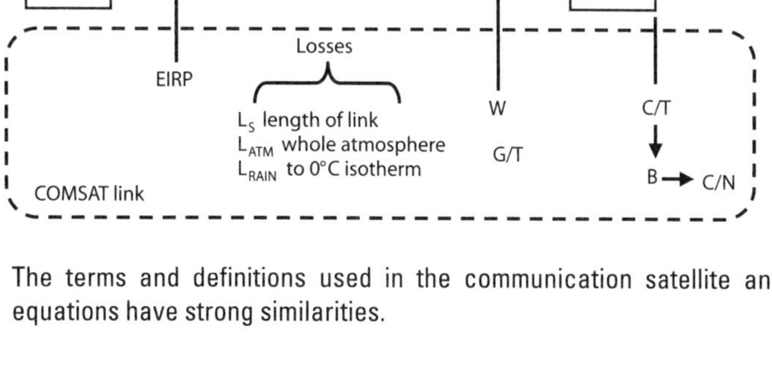

Figure 7.15 The terms and definitions used in the communication satellite and EW link equations have strong similarities.

7.7 Relating Communication Satellite and EW Forms of Equations

Figure 7.15 shows a link with both the terrestrial and communication satellite link values. As you can see, they have strong similarities, but there are more steps in the calculations made for communication satellites.

The terrestrial link definitions that we use in most EW applications have the assumption that all components and the angular area falling within the antenna's beam are at 290K.

In terrestrial links, we speak of the ERP leaving the transmit antenna as the product (or sum in decibels) of the transmitter power and the antenna gain. There is room for confusion in this concept, as there is an often unspoken assumption that the peak gain of antenna pattern is aimed at the receiver. In EW, this is often not the case, so we must understand that the ERP in an EW link must include the antenna gain in the direction of the receiver. The COMSAT link uses EIRP, which defines the power that would have to be input to an isotropic (unity gain) antenna to cause the effective radiated power at the peak of the beam. The antenna pointing error is handled separately.

The link losses are different only in the ranges over which they apply. In terrestrial links atmospheric loss per kilometer is normally considered constant over the whole link path, while for COMSAT links, the path goes through the whole atmosphere so there is a fixed-loss amount for any given frequency and elevation angle.

Rain loss for terrestrial links depends on the density of the rain and a model of the rain density profile along the link path, while SATCOM links are subject to rain only from the 0°C isotherm to the ground station along the link path. The spreading losses for both cases use the line-of-sight propagation model. However, for terrestrial links below microwave, different propagation loss models can apply.

For terrestrial links, it is sometimes useful to define the power arriving at the location of the receiver. This is either defined in terms of the received field strength (in microvolts per meter) or in dBm (using the artifice of the power that would be produced by an ideal isotropic antenna at that point). For COMSAT links, we define the illumination level "W" in dbW/m^2.

From this point, the terms become different. We define the received power in a terrestrial link as the power leaving the antenna and input to the receiving system. We define the sensitivity as the product of kTB in the effective receiver bandwidth, the noise figure of the receiver, and the required predetection SNR (commonly called C/N, but also called the RFSNR). The quality of the demodulated output from the receiver is described in terms of the SNR which is usually in decibels.

For the sitcom link, we apply a receiver quality factor to the illumination level. This quality factor is the gain of the antenna divided by the receiving system noise temperature. This allows us to calculate the carrier to noise temperature ratio which is independent of the receiver bandwidth. Then, the bandwidth is applied to develop the carrier-to-noise ratio (C/N).

7.8 Jamming of Satellite Links

In EW, we are concerned both with the vulnerability of friendly SATCOM links to jamming and the jamming of enemy SATCOM links. For convenience, we will take the point of view of the jammer in this discussion.

Like any other type of jamming, it is necessary to jam the receiver—not the transmitter. This confusion arises because radars have collocated transmitters and receivers. SATCOM links are at the other extreme, since their receivers and transmitters are far removed from each other. Since most satellite links are bidirectional, location of the transmitters can tell us the locations of the (nonemitting) receivers. This is important information, because the distances involved, require directional jamming antennas in most cases.

Note that COMSAT signals almost always use digital modulations, so the discussions in Chapter 5 about jamming digital signals apply.

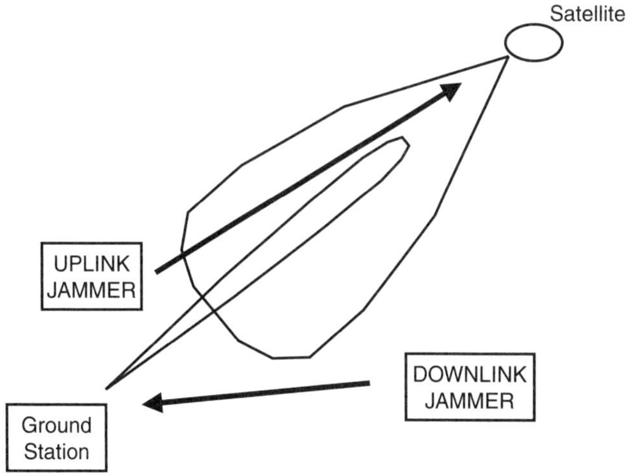

Figure 7.16 COMSAT jamming vulnerability is a strong function of the jamming geometry.

7.8.1 Downlink Jamming

Figure 7.16 shows the SATCOM jamming geometry considerations. First, let's jam the downlink (from the satellite to the ground station). The ground station antenna in most cases has a fairly narrow directional antenna. Therefore, we must either be very close to the ground station or have sufficient jamming power to achieve adequate J/S through the side lobes of the antenna—which are likely to be very low. The jamming power in the receiver must be adequate to cause sufficient bit errors. If the jammer must be distant from the ground station, this can require a great deal of jamming power. It is to be expected that the downlink will also have some level of spectrum spreading modulation for A/J protection, and may also have error correction coding. Both of these features increase the amount of jamming power required to create adequate bit error density for effective jamming. The compensating factor is that signals from satellites are likely to be quite low level because of the spreading loss.

There are two important cases that have different considerations: satellite cell phone jamming and GPS jamming.

For logistical reasons, satellite cell phones can be expected to have fairly omnidirectional antenna patterns. Narrowbeam antennas are ungainly and must be oriented toward the satellite. Because of the spreading losses from synchronous satellites, cell phones can be expected to operate with low-Earth-orbit satellites, making satellite tracking impractical. This means that the jammer can expect to see the same receiving antenna gain as that toward the satellite. The jammer can, however, use a directional antenna to optimize its power toward the receiver location. Therefore, antijam (A/J) protection from spectrum spreading and error correction codes are the only practical electronic protection (EP) measures for satellite cell phones.

GPS is not a SATCOM program, but it is an important EW consideration worthy of discussion here. The received GPS signals are very weak, on the order of –150 dBm, so it is an easy matter to generate adequate jamming signals if the jammer can be within line of sight. The GPS signal has two levels of spectrum spreading: publicly available CA code and highly restricted access P code. CA code signals have about 40 dB of A/J protection using open codes, which still allows jamming with relatively weak signals. It is commonly said that you can jam CA code with "a penny and a lemon peel," but this only applies to situations in which the jammer has good line of sight.

P code signals have an additional level of spectrum spreading and use secure codes, so they have an additional 40 dB of A/J protection. Thus, a jamming signal must have enough power to overcome 80 dB of A/J protection and still create adequate J/S.

7.8.2 Uplink Jamming

Jamming a SATCOM uplink is geometrically less challenging than jamming the downlink, because the receiving antenna in the satellite is pointing at the Earth. For a synchronous satellite with an Earth coverage antenna, the jammer can be anywhere on about 45% of the Earth's surface and still jam into the main lobe. Even narrow-beam antennas cover a great deal of real estate, from synchronous or low-Earth-orbit satellites, so only specrum spreading and error correction codes are dependable EP measures. Still, if the downlink uses a narrow-beam antenna and if the jammer cannot be within the Earth footprint of that antenna, the jammer must overcome antenna sidelobe isolation in addition to the A/J features of the uplink.

In all cases, there is a great deal of spreading loss that must be over come to jam a satellite uplink because of the distance to the satellite. This is balanced by the fact that the uplink transmitter must propagate the same distance. The jammer effective radiated power must therefore be greater than the uplink transmitter power by the amount of J/S required, the A/J protection factor, and the antenna isolation if applicable.

Appendix A: Problems with Solutions

In response to many requests, this appendix comprises problems from both the subject matter in the *EW 101* and *EW 102* books. Each of these problems is worked to 1-dB resolution, using the corresponding forms of the formulas. Where it is appropriate to use a nomograph or graph from the text, it is copied and its use to solve the problem is shown.

When antenna gains are given, they refer to the gain in dBi in the direction of the receiver (for transmit antennas) and in the direction of the transmitter (for receiving antennas) unless otherwise stated in the problem.

In these problems and solutions, "log" is taken to mean "$\log_{10}$."

Remember that decibel formulas require that inputs be made in the proper units. These units are stated along with the formulas in the referenced sections.

Part 1 Problems from the *EW 101* Book

Each of these problems is from material in *EW 101*. The section numbers listed are the sections of the *EW 101* book where the formulas and explanations are found.

Problem 101–1: Convert 4W into dBm.

►(Section 2.1.2)

$$4\text{W} / 1\text{ mW} = 4{,}000$$

$$10 \log_{10}(4{,}000) = 36 \text{ dBm}$$

Problem 101–2: Convert 70 dBW to dBm.

►(Section 2.1.2)

$$1\text{W} = 1{,}000 \text{ mW}$$

$$10 \log (1{,}000) = 30 \text{ dB}$$

$$70 \text{ dBW} + 30 \text{ dB} = 100 \text{ dBm}$$

Problem 101–3: Calculate the line-of-sight spreading loss for a 1-GHz signal at 50 km.

►(Section 2.2.2)

$$\begin{aligned} L_S &= 32 + 20\log(d) + 20\log(F) \\ &= 32 + 20\log(50) + 20\log(1{,}000) \\ &= 32 + 34 + 60 = -126 \text{ dB} \end{aligned}$$

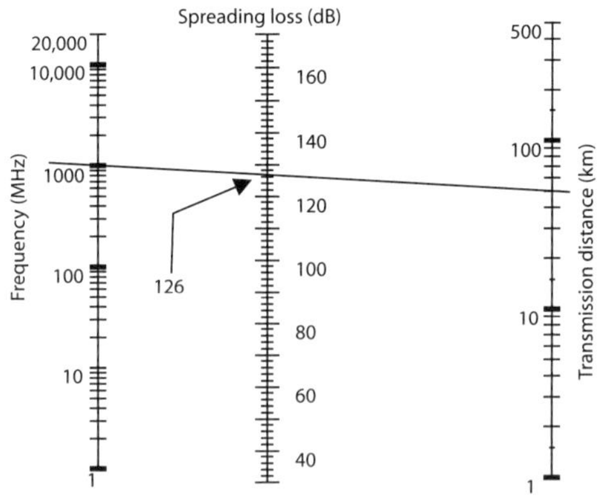

Problem 101–3

Alternately, use the nomograph in Figure 2.2:

Draw a line from the frequency (in megahertz) to the distance (in kilometers).

Line crosses center scale (spreading loss in decibels) at 126 dB.

Problem 101–4: Find the atmospheric attenuation for a 10-GHz signal at 20 km.

►(Section 2.2.2)

For this problem, you need to use the graph in Figure 2.3.

Start with the frequency (15 GHz) on the abscissa of the graph. Note that this is a logarithmic scale, so 15 is about 0.7 of the way between 10 and 20. Go up to the curve, then left to the ordinate. Read 0.04 dB per kilometer of range.

Since the range is 20 km, the atmospheric attenuation is 0.04 × 20 = 0.8 dB.

Problem 101–5: Calculate the received signal strength from a 2-GHz, 10-W transmitter with antenna gain of 10 dB toward the receiver, which is 27 km away. The receiving antenna gain (toward the transmitter) is 20 dB.

►(Section 2.2.1)

$$10\text{W} = 10{,}000 \text{ mw}$$

$$10 \log (10{,}000 \text{ mw}) = +40 \text{ dBm}$$

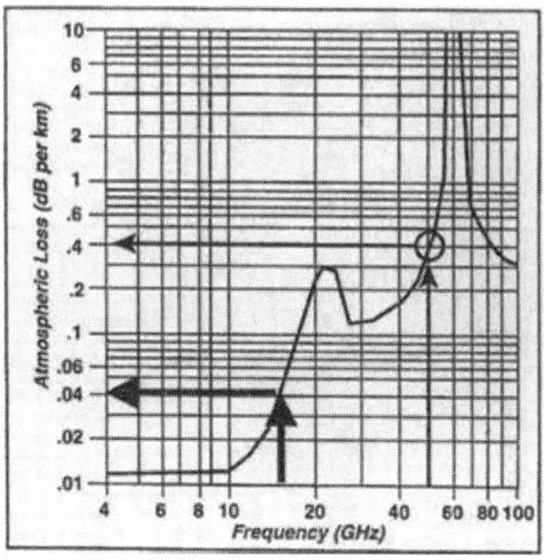

Problem 101–4

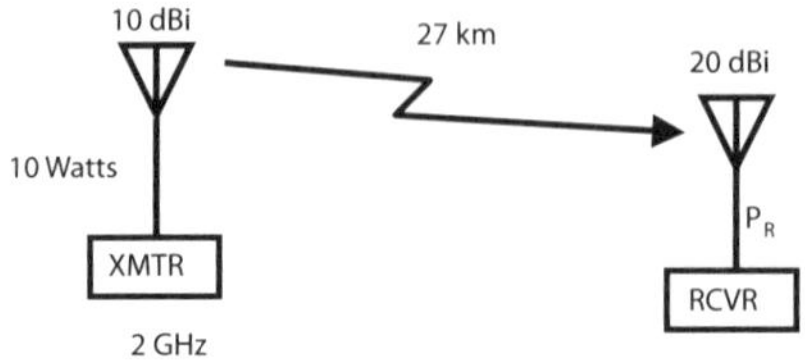

Problem 101–5

$$\begin{aligned} P_R &= P_T + G_T - 32 - 20\log(d) - 20\log(F) + G_R \\ &= +40 + 10 - 32 - 20\log(27) - 20\log(10{,}000) + 20 \\ &= 40 + 10 - 32 - 29 - 80 + 20 = -71 \text{ dBm} \end{aligned}$$

Problem 101–6: Find range at which a receiver with a sensitivity of –80 dBm and a –10-dBi antenna gain (toward the transmitter) can receive a signal from a 5-GHz signal with 100-kW transmitter power and antenna gain of 10 dB toward the receiver.

►(Section 2.2.4)

$$100 \text{ kW} = 100{,}000{,}000 \text{ mw}$$

$$10\log(100{,}000{,}000) = +80 \text{ dBm}$$

$$\text{Set } P_R = \text{Sens} = P_T + G_T - 32 - 20\log(d) - 20\log(F) + G_R$$

$$\begin{aligned} \text{Solve for } 20\log(d) &= P_T + G_T - 32 - 20\log(F) + G_R - \text{Sens} \\ &= +80 + 10 - 32 - 20\log(5{,}000) + (-10) - (-80) \\ &= +80 + 10 - 32 - 74 - 10 + 80 \\ &= 54 \end{aligned}$$

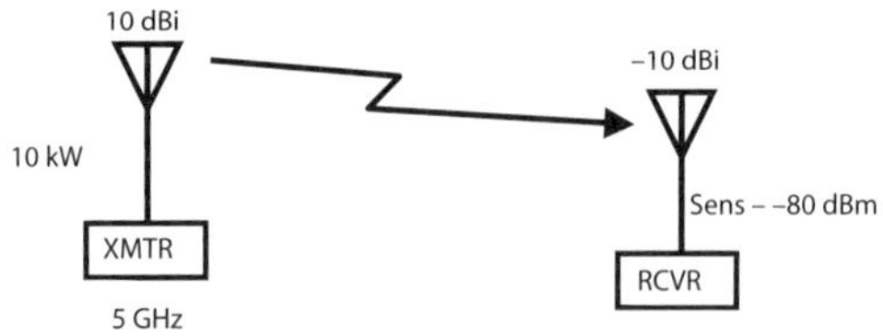

Problem 101–6

$$d = \text{antilog}(54\ /\ 20) = \text{antilog}(2.7) = 501 \text{ km}$$

Problem 101–7: Find the sensitivity in dBm of a receiver that is specified at 1 μv/m at 100 MHz.

►(Section 2.3.2)

$$\begin{aligned} P &= -77 + 20\log(E) - 20\log(F) \\ &= -77 + 20\log(1) - 20\log(100) \\ &= -77 + 0 - 40 = -117 \text{ dBm} \end{aligned}$$

Problem 101–8: Find the sensitivity in microvolts per meter of a receiver specified as –100 dBm sensitivity at 50 MHz.

$$\begin{aligned} E &= \text{antilog}\{[P + 77 + 20\log(F)]\ /\ 20\} \\ &= \text{antilog}\{[-100 + 77 + 34]\ /\ 20\} \\ &= \text{antilog}\{0.55\} = 3.5\ \mu\text{v/m} \end{aligned}$$

Problem 101–9: What is the received power in a radar receiver if the transmitter power is 10 kw, the antenna gain is 30 dBi, the frequency is 10 GHz, the target is 25 km away, and the target's RCS is 20 m^2?

►(Section 2.3.3)

$$P_R = P_T + 2G - 103 - 40\log(D) - 20\log(F) + 10\log(\text{RCS})$$

$$P_T = 10 \text{ kw} \quad 10\log(100{,}000{,}000) = +70 \text{ dBm}$$

$$40\log(D) = 40\log(25) = 56$$

$$20\log(F) = 20\log(10{,}000) = 80$$

$$10\log(\text{RCS}) = 10\log(20) = 13$$

$$P_R = 70 + 60 - 103 - 56 - 80 + 13 = -96 \text{ dBm}$$

Problem 101–10: What is the FZ distance for a 100-MHz signal from a transmitter 2m above the terrain and a receiver 1,000m above the terrain?

►(Section 2.3.5)

$$\begin{aligned} FZ &= (h_T \times h_R \times f) / 24{,}000 \\ &= 2 \times 1{,}000 \times 100 / 24{,}000 = 8.3 \text{ km} \end{aligned}$$

Please note that there is an error in some *EW 101* printings in this formula. The book says to divide by 75,000—the correct division factor (rounded) is 24,000.

The other formula for FZ is:

$$FZ = 4\pi \times h_T \times h_R / \lambda$$

The wavelength of a 100-MHz signal is 3×10^8 m/s/10^8 Hz = 3m

$$4\pi \times 2 \times 1{,}000 / 3 = 8{,}377\text{m}$$

which is more accurate because 24,000 has been rounded for convenience.

Problem 101–11: Determine the spreading loss for a 100-MHz signal from a 2-m-high transmit antenna to a 1,000m-high receiving antenna at a distance of 25 km—using the two-ray model.

►(Section 2.3.5)

Note that the two-ray model is appropriate because (as calculated in Problem 101–10) the FZ distance is less than the transmission range.

$$\begin{aligned} L_S &= 120 + 40\log(d) - 20\log(h_T) - 20\log(h_R) \\ &= 120 + 40\log(25) - 20\log(2) - 20\log(1{,}000) \\ &= 120 + 56 - 6 - 60 = 110 \text{ dB} \end{aligned}$$

Problem 101–12: Determine the received power in a receiver with a 2-dB antenna gain and 100-m antenna height, 20 km from a 1-W, 50-MHz transmitter with a 2-dB antenna gain and 2-m antenna height.

►(Section 2.3.5)

$$FZ = 2 \times 100 \times 50 / 24{,}000 = 0.417 \text{ km}$$

The link is longer than the FZ distance, so the two-ray model should be used.

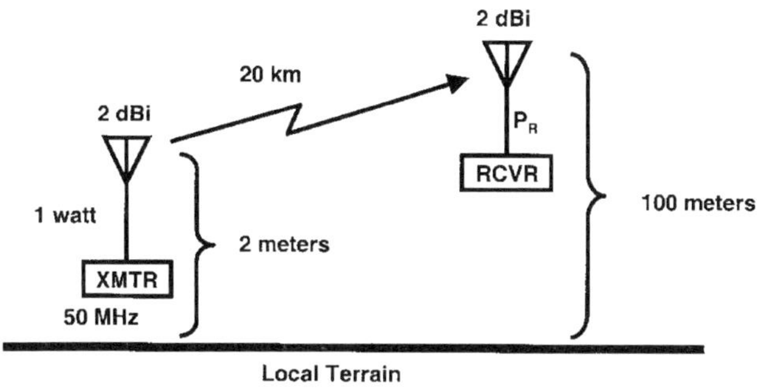

Problem 101–12

$$P_R = P_T + G_T - [120 + 40\log(d) - 20\log(h_T) - 20\log(h_R)] + G_R$$
$$= +30\text{ dBm} + 2\text{ dB} - 120 - 40\log(20) + 20\log(2) + 20\log(100) + 2\text{ dB}$$
$$= 30 + 2 - 120 - 52 + 6 + 40 + 2 = -92\text{ dBm}$$

Problem 101–13: In this spherical triangle (where capital letters indicate angles and a lower case letter is the side opposite the angle indicated by the corresponding capital letter), if *a* is 35°, *A* is 42°, and *B* is 52°, what is the value of *b*?

►(Section 2.4.3)

In any spherical triangle, $\sin(a)/\sin(A) = \sin(b)/\sin(B) = \sin(c)/\sin(C)$

$$\sin(b) = \sin(a)\times\sin(B)/\sin(A)$$
$$b = \arcsin[\sin(a)\times\sin(B)/\sin(A)]$$
$$= \arcsin[0.574\times 0.788/0.669] = \arcsin[0.676] = 42.5°$$

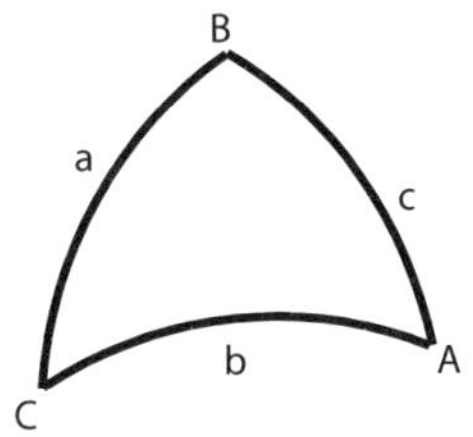

Problem 101–13

Problem 101–14: In the same spherical triangle, find a if b is 37°, c is 45°, and A is 67°.

►(Section 2.4.3)

$$\cos(a) = \cos(b) \times \cos(c) + \sin(b) \times \sin(c) \times \cos(A)$$
$$a = \arccos[\cos(b) \times \cos(c) + \sin(b) \times \sin(c) \times \cos(A)]$$
$$= \arccos[0.799 \times 0.707 + 0.602 \times 0.707 \times 0.391] = \arccos[0.731] = 43.0°$$

Problem 101–15: In the same spherical triangle, find C if A is 120°, B is 35°, and c is 50°.

►(Section 2.4.3)

$$\cos(C) = -\cos(A) \times \cos(B) + \sin(A) \times \sin(B) \times \cos(C)$$
$$C = \arccos[-\cos(A) \times \cos(B) + \sin(A) \times \sin(B) \times \cos(C)]$$
$$= \arccos[0.5 \times 0.819 + 0.866 \times 0.574 \times 0.643]$$
$$= \arccos[0.729] = 43.2°$$

Problem 101–16: In a right spherical triangle with c the side opposite the 90° angle, the two sides being a and b, and A and B the angles opposite the corresponding sides, what is c if a is 47° and b is 85°?

►(Section 2.4.4)

$$\cos(c) = \cos(a) \times \cos(b)$$
$$c = \arccos[\cos(a) \times \cos(b)]$$
$$= \arccos[0.682 \times 0.087] = \arccos[0.059] = 86.6°$$

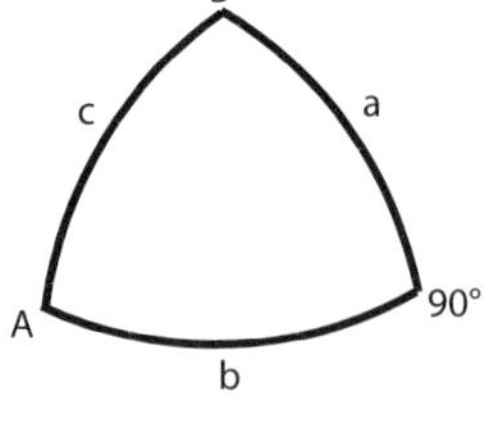

Problem 101–16

Problem 101–17: In the same triangle, find c if A is 80° and b is 44°.

►(Section 2.4.4)

$$\cos(A) = \tan(b) \times \operatorname{ctn}(c)$$
$$\operatorname{ctn}(c) = \cos(A) / \tan(b)$$

Note that ctn = 1/ tan (nice because the calculator does not have a ctn function).

$$\text{So } \tan(c) = \tan(b) / \cos(A)$$
$$c = \arctan[\tan(b) / \cos(A)]$$
$$= \arctan[0.966 / (0.174)] = \arctan[5.563] = 79.8°$$

Problem 101–18: The gain of a radar warning receiver quadrant antenna is reduced from boresight gain by 0.2 dB per degree (out to 90°) from the antenna boresight. The elevation of an emitter relative to the antenna boresight is 42° and the azimuth relative to the boresight is 65°. What is the antenna gain toward that emitter relative to the antenna boresight gain?

►(Section 2.4.4)

First, use Napier's rules for a right spherical triangle to find the spherical angle from boresight to the emitter.

$$\text{Spherical angle} = \arccos[\cos(Az) \times \cos(El)]$$
$$= \arccos[0.423 \times 743] = \arccos[0.314] = 71.7°$$

The gain reduction from boresight is 71.7° × 0.2 dB/degree = 14.3 dB.

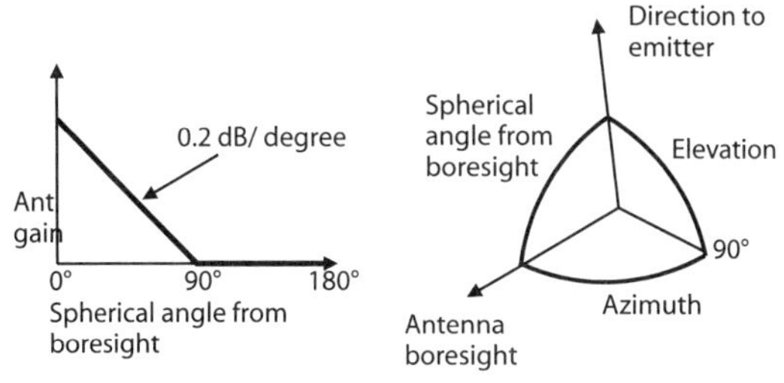

Problem 101–18

Problem 101–19: Determine the Doppler shift in a signal transmitted from a moving transmitter to a fixed receiver. The transmitter location is 5 km north, 7 km east, 1 km up. The receiver location is 25 km north, 15 km east, 2 km up. The transmitter's velocity vector is 150 m/s at 20° elevation angle, 5° azimuth angle. The transmitted frequency is 10 GHz.

►(Section 2.5.2)

Note that east is in the positive *X*-direction, north is positive *Y*, and *Z* is up.

$$AZ_R = \arctan\left[\left(X_R - X_T\right) / \left(Y_R - Y_T\right)\right] = \arctan\left[(15-7)/(25-5)\right]$$
$$= \arctan[0.4] = 21.8°$$
$$El_R = \arctan\left\{\left(Z_R - Z_T\right) / \text{sqrt}\left[\left(X_R - X_T\right)^2 + \left(Y_R - Y_T\right)^2\right]\right\}$$
$$= \arctan\left\{(2-1)/\text{sqrt}\left[(15-7)^2 + (25-5)^2\right]\right\} = \arctan\{2/21.5\}$$
$$= \arctan\{0.093\} = 5.3°$$

Then, using Figure 2.14:
The spherical angle from the velocity vector to north is

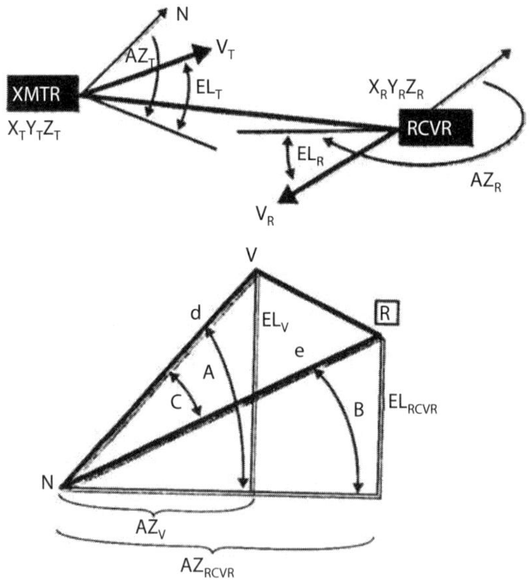

Problem 101–19

$$\cos(d) = \cos(Az_V) \times \cos(EL_V)$$
$$d = \arccos[\cos(Az_V) \times \cos(EL_V)] = \arccos[\cos(5°) \times \cos(20°)]$$
$$= \arccos[0.996 \times 0.940] = \arccos[0.936] = 20.6°$$

$$e = \arccos[\cos(Az_R) \times \cos(El_R)] = \arccos[\cos(21.8°) \times \cos(5.3°)]$$
$$= \arccos[0.925] = 22.4°$$

The angles A and B are found from:

$$A = \text{arc ctn}[\sin(Az_R) / \tan(El_V)] = \text{arc ctn}[\sin(5°) / \tan(20°)]$$
$$= \text{arc ctn}[0.087 / 0.364] = \text{arc ctn}[0.239] = \text{arc ctn}[1 / 0.239]$$
$$= \text{arc ctn}[4.18] = 76.5°$$

$$B = \text{arc ctn}[\sin(Az_R) / \tan(EL_R)] = \text{arc ctn}[\sin(21.8°) / \tan(5.3)]$$
$$= \text{arc ctn}[0.371 / 0.0928] = \text{arc ctn}[4.0°] = \arctan[0.250] = 14.0°$$

Angle C is $A - B = 76.5° - 14.0° = 62.5°$.

Then, the spherical angle from the velocity vector to the receiver is found from

$$VR = \arccos[\cos(d) \times \cos(e) + \sin(d) \times \sin(e) \times \cos(C)]$$
$$= \arccos[\cos(20.6°) \times \cos(22.4°) + \sin(20.6°) \times \sin(22.4°) \times \cos(62.5°)]$$
$$= \arccos[0.936 \times 0.925 + 0.352 \times 0.381 \times 0.462] = \arccos[0.928] = 21.9°$$

The rate of change of range between the transmitter and the receiver is:

$$V_{REL} = V \cos(VR) = (150 \text{ m/s}) / \times 0.928 = 139.2 \text{ m/s}$$

The Doppler frequency shift is:

$$\Delta f = f \times V_{REL} / c = 10^{10} \text{ Hz} \times 139.2 \text{ m/s} / 3 \times 10^{8} \text{ m/s} = 4{,}640 \text{ Hz}$$

Problem 101–20: What is the effective area of an antenna with boresight gain of 30 dB at 1 GHz?

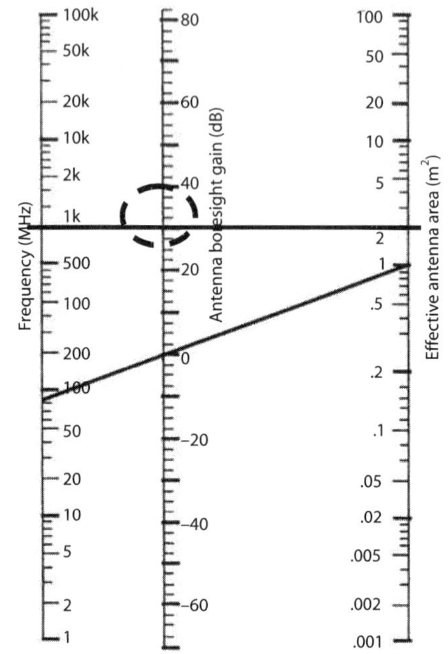

Problem 101–20

►(Section 3.3.2)

Use the nomograph of Figure 3.5. Draw a line from 1K (1,000 MHz) through 30 dB on the center scale and read the effective area in square meters on the right-hand scale.

2 m^2

Problem 101–21: What is the gain of a 55% efficient parabolic dish antenna with 2m diameter operating at 5 GHz?

►(Section 3.3.3)

Use the nomograph of Figure 3.6. Draw a line from 5 GHz on the left-hand scale to 2m on the right-hand scale. Read the gain on the center scale.

Approximately 38 dB.

Problem 101–22: What is the gain of a 55% efficient parabolic dish antenna with a 3-dB beamwidth of 10° in elevation and 25° azimuth?

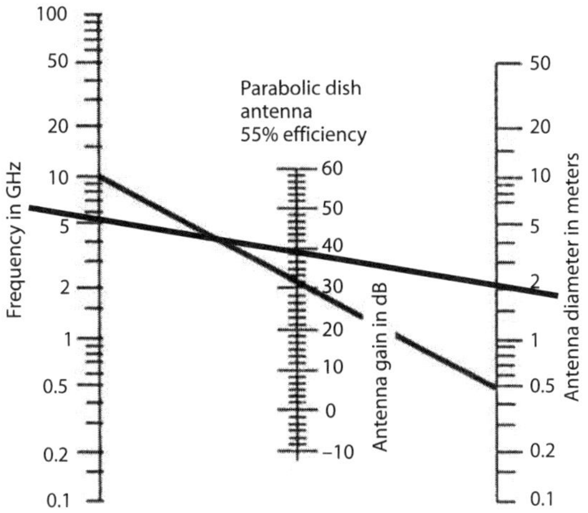

Problem 101–21

►(Section 3.3.4)

$$\text{Gain}\left(\text{not in decibels}\right) = 29{,}000 \,/\, \left(10 \times 25\right) = 116$$

$$\text{Gain}\left(\text{in decibels}\right) = 10\log\left(116\right) = 20.6 \text{ dB}$$

Problem 101–23: What is the sensitivity of a receiver with 10-MHz bandwidth, Noise figure = 5 dB, and required signal-to-noise ratio = 13 dB?

►(Section 4.11.2)

$$\text{kTB}(\text{dBm}) = -114 + 10\log\left(\text{Bandwidth}/1\text{ MHz}\right) = -114 + 10\log(10)$$
$$= -114 + 10 = -104 \text{ dBm}$$

$$\text{Sensitivity} = \text{kTB}(\text{dBm}) + \text{NF}(\text{dB}) + \text{SNR}(\text{dB})$$
$$= -104 \text{ dBm} + 5 \text{ dB} + 13 \text{ dB} = -86 \text{ dBm}$$

Problem 101–24: What is the noise figure of a receiver system with a 3-dB noise figure, 25-dB gain preamplifier. Loss before the preamplifier = 1 dB. Loss between preamplifier and receiver = 13 dB. Receiver noise figure = 10 dB?

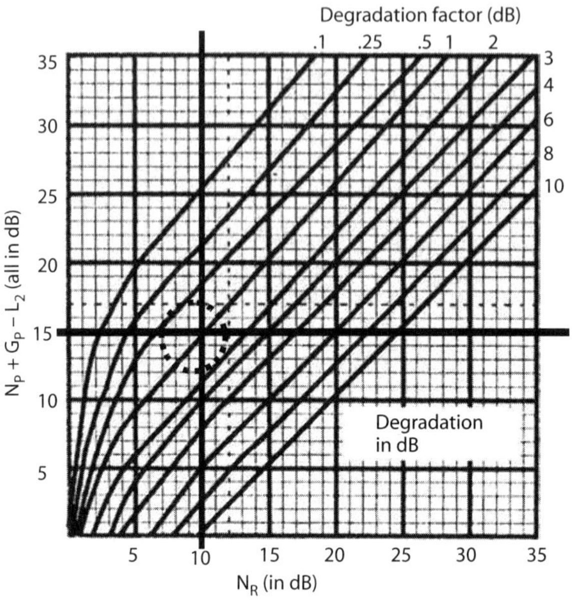

Problem 101–24

►(Section 4.11.2.2)

Use the graph in Figure 4.17.

Find the preamp gain + the preamp noise figure–the loss before the receiver on the ordinate: 25 + 3 – 13 = 15. Draw a horizontal line through this point.

Draw a vertical line through the receiver noise figure on the abscissa.

These two lines cross on the line indicating the degradation (1 dB in this case).

System noise figure = loss before preamp + preamp noise figure + degradation = 1 dB + 3 dB + 1 dB = 5 dB

Problem 101–25: What postdiscrimination signal-to-noise ratio is achieved with a phase locked loop discriminator receiving an FM signal with modulation index 5 and predetection signal-to-noise ratio 4 dB?

►(Section 4.12)

The FM improvement factor for a signal above threshold is:

$$IF_{FM}(\text{dB}) = 5 + 20\log(\text{mod index}) = 5 + 20\log(5) = 5 + 14 = 19$$

The postdetection SNR is 4 dB + 19 dB = 23 dB.

Problem 101–26: What is the postdetection signal-to-noise ratio (actually the signal-to-quantization ratio) for a signal that is digitized with 5 bits per sample?

►(Section 4.13.1)

$$\text{SQR}(\text{dB}) = 5 + 3(2m - 1)$$

$$m = \text{bits/sample} = 5$$

$$\text{SQR}(\text{dB}) = 5 + 3(9) = 32\text{ dB}$$

Problem 101–27: What is the bit error rate for a coherent PSK modulated signal which has an E_b/N_0 of 8 dB?

►(Section 4.13.2)

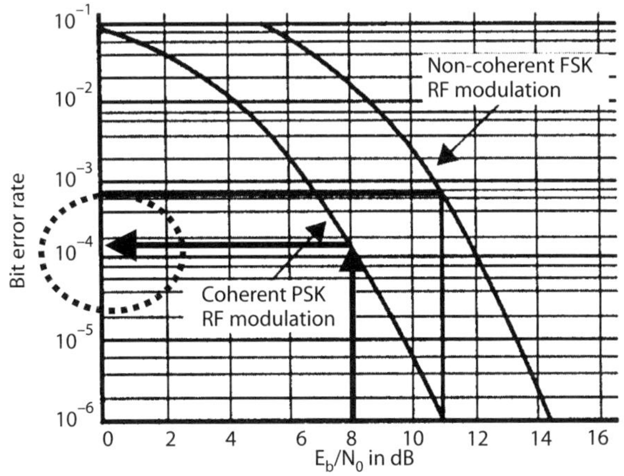

Problem 101–27

Use the graph in Figure 4.20. Please note that the abscissa of this graph should be labeled E_b/N_0 (in decibels).

Draw a vertical line from 8 dB on the abscissa up to the coherent PSK curve, then left to the ordinate and read about 1.2×10^{-4} bit error rate.

Problem 101–28: What is the probability of intercept within 1 second for a radar signal with 2-μs pulse width, 1,000-pps PRF, between 2 and 4 GHz. It has a beamwidth of 5° and a circular scan of a 5-second period. Receiver sensitivity is adequate to see the radar's whole 3-dB beamwidth, and receiver bandwidth is 10 MHz.

►(Section 6.2)

The 1-second allowable time starts with the arrival of the first pulse at the receiving antenna, so we must find the signal during the first beam.

The beamwidth is 5°, and the threat antenna covers 360° in 5 seconds, so the beam duration is 5 sec (5/360) = 69.4 ms. This means we can receive 69 pulses as the beam scans past the receiver.

If we step tune at maximum rate, our bandwidth must stay at one frequency for time = 1/bandwidth or 100 ns. The pulse is 2 μs long, so we can look 20 times during the pulse (covering 20 × 10 MHz = 200 MHz). The probability of intercepting each pulse is

$$200 \text{ MHz} / 2{,}000 \text{ MHz} = 10\%$$

If we step one bandwidth during each pulse interval (1 ms) for 69 intervals, we will cover 69 × 10 MHz = 690 MHz. The probability of intercepting one of the 69 pulses is

$$690 \text{ MHz} / 2000 \text{ MHz} = 34.5\%$$

Note: The following is really not fair, because it is beyond the scope of the text, but there is another solution. That is to find the probability that we get one or more pulses if we sweep as fast as possible and consider that we get 69 tries. The probability of at least 1 success in 69 tries is:

$$1(1-p)^{69}$$

where p is the probability of success in one try. Since p = 10%, the probability of success in 69 tries is $1 - 0.9^{69}$ = 0.9993 or 99.93%.

Problem 101–29: Find the maximum line-of-sight distance between a transmitter using a 100-m-high antenna and a receiver on an aircraft at 2,000-m altitude. Assume smooth 4/3 Earth.

►(Section 6.3.2)

The maximum line of sight distance is

$$D = 4.11 \times \left[\text{sqrt}(H_T) + \text{sqrt}(H_R)\right]$$

D is in kilometers, H values are in meters

$$= 4.11 \times (10 + 44.7) = 224.8 \text{ km}$$

Problem 101–30: Calculate the J/S achieved by a 100W jammer with a 10-dB antenna, 30 km from a receiver with a clear line-of-sight path—when the desired transmitter has 1-W ERP and is 10 km from the receiver, and the receiver has a whip antenna.

►(Section 9.2.3)

$$\text{J/S} = P_J - P_T + G_J - G_T - 20\log(D_J) + 20\log(D_S) + G_{RJ} - G_R$$

The sum of P_T and G_T is the 1-W ERP of the transmitter = 30 dBm. G_{RJ} and G_R are equal, so these two terms cancel each other. So:

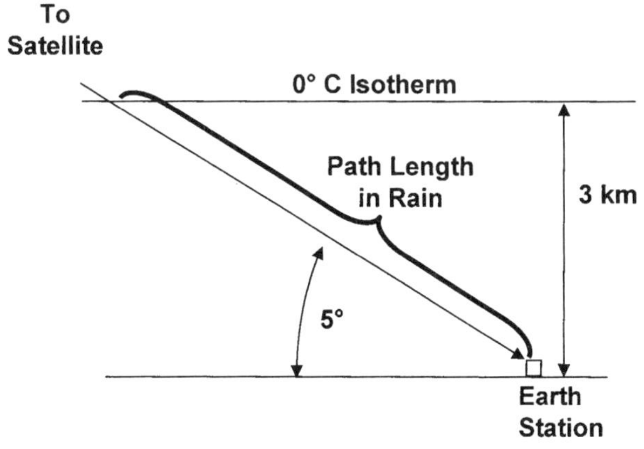

Problem 101–30

$$J/S = P_J + G_J - ERP_T - 20\log(D_J) + 20\log(D_S)$$
$$= +50 + 10 - 30 - 20\log(30) + 20\log(10) = 50 + 10 - 30 - 30 + 20 = 20 \text{ dB}$$

Problem 101–31: Calculate the J/S achieved by a 1-kW ERP self-protection jammer 15 km from a radar with 1-MW ERP. The target has a radar cross section of 2 m^2.

▸(Section 9.2.3)

$$J/S = 71 + P_J - P_T + G_J - G_R + 20\log(D_T) - 10\log(RCS)$$

$P_J + G_J = ERP_J$ and $P_T + G_T = ERP_R$ and 1 kw = +60 dBm and 1 MW = + 90 dBm

So:

$$J/S = 71 + ERP_J - ERP_R + 20\log(D_T) - 10\log(RCS)$$
$$= 71 + 60 - 90 + 20\log(15) - 10\log(2) = 71 + 60 - 90 + 24 - 3 = 62 \text{ dB}$$

Problem 101–32: Calculate the burn-through range for the above radar, jammer and target if burn through occurs at 0-dB J/S.

▸(Section 9.3.5)

First, solve the J/S equation for 20 log (D_T)

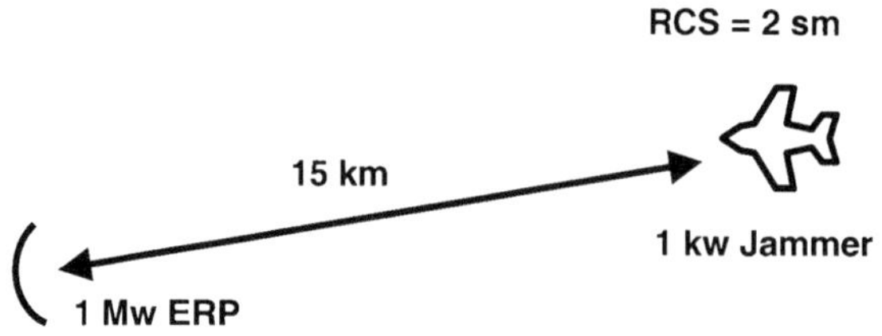

Problem 101–31

$$20\log(D_T) = -71 - \text{ERP}_\text{J} + \text{ERP}_\text{R} + 10\log(\text{RCS}) + \text{J/S}$$

$$= -71 - 60 + 90 + 2 + 0 = -39$$

$$D_T = \text{antilog}[-39/20] = 0.0112 \text{ km} = 112\text{m}$$

Problem 101–33: Calculate the J/S achieved by a standoff jammer with 2-kW transmitter power and 18-dB antenna gain at 25-km range from the radar. The radar ERP is 1 Mw including 30-dBi main-beam antenna gain. The standoff jammer is in a 0-dBi sidelobe. The target aircraft is 10 km from the radar and has RCS of 2 m^2.

►(Section 9.3.4)

$$\text{J/S} = 71 + P_J - P_T + G_J + G_{RJ} - 2G_R - 20\log(D_J) + 40\log(D_T) - 10\log(\text{RCS})$$

$$1 \text{ Mw} = +90 \text{ dBm} \quad \text{and} \quad 2 \text{ kw} = +63 \text{ dBm}$$

$$P_T = \text{ERP}_\text{R} - G_R = +90 \text{ dBm} - 30 \text{ dB} = +60 \text{ dBm}$$

$$\text{J/S} = 71 + 63 - 60 + 18 + 0 - 2(30) - 20\log(25) + 40\log(10) - 10\log(2)$$
$$= 71 + 63 - 60 + 18 + 0 - 60 - 28 + 40 - 3 = 41 \text{ dB}$$

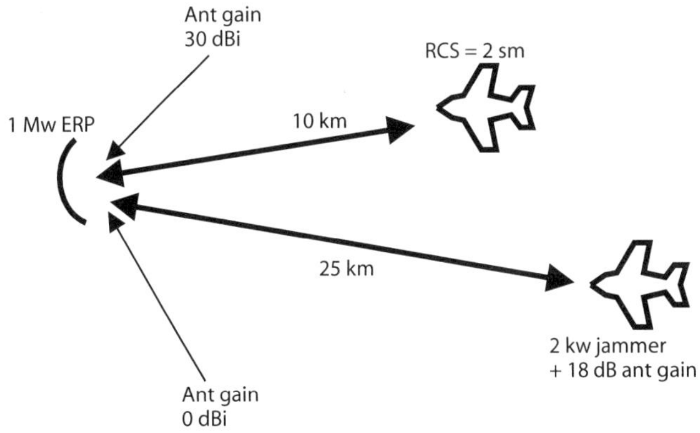

Problem 101–33

Problem 101–34: Calculate the burn-through range for the above radar, target, and standoff jammer.

►(Section 9.3.4)

Note that the burn-through range refers to the range from the radar to the target. The standoff jammer stays at constant range from the radar. The required J/S is 0 dB.

Solve the standoff jamming J/S equation for 40 log (D_T):

$$40 \log(D_T) = -71 - P_J + P_T - G_J - G_{RJ} + 2G_R + 20 \log(D_J) + 10 \log (\text{RCS}) + \text{J/S} = -71 - 63 + 60 - 18 - 0 + 60 + 28 + 3 + 0 = -1$$

$$D_T = \text{antilog}\,[-1/40] = 0.944 \text{ km} = 944\text{m}$$

Problem 101–35: What are the dimensions of the radar resolution cell if the range to the target is 15 km, the 3-dB beamwidth is 2°, and the pulse width is 2 μs?

►(Section 9.9.2)

The depth of the cell is: 0.5 (c × PW) = 0.5 × 3 × 10^8 × 2 × 10^{-6} = 300m.

The width of the cell is: 2 R[sin(BW/2)] = 2 × 15,000 × sin(1°) = 2 × 15,000 × 0.0175 = 524m

Problem 101–36: What RCS is simulated by an active decoy that receives a 5-GHz signal at –30 dBm and transmits a 1-kW return signal?

►(Section 10.7.3)

$$1 \text{ kw is } + 60 \text{ dBm}$$

$$\text{The effective gain is } +60 \text{ dBm} - (-30 \text{ dBm}) = 90 \text{ dB}$$

$$\text{RCS (dBsm)} = 39 + G - 20 \log(F) = 39 + 90 - 20 \log(5{,}000) = 39 + 90 - 74 = 55 \text{ dBsm}$$

$$\text{RCS}(\text{m}^2) = \text{antilog}\,(55\,\text{dBsm} / 10) = 316{,}000 \text{ m}^2$$

Part 2 Problems from the *EW 102* Book

Each of these problems is from material in this book. The section numbers listed are the sections of the book where the formulas and explanations are found.

Problem 102–1: Determine the maximum unambiguous range of a radar with a pulse repetition frequency of 5,000 pulses per second.

►(Section 2.5.1)

$$\text{PRI} = 1/\text{PRF} = 1/5{,}000 \text{ per sec} = 200\,\mu\text{s}$$

$$R_{\text{MAX}} = 0.5 \times \text{PRI} \times c = 0.5 \times 2 \times 10^{-4} \text{ sec} \times 3 \times 10^{8} \text{ m/s} = 30{,}000\text{m}$$

$$= 30 \text{ km}$$

Problem 102–2: Determine the minimum range for a radar with a pulse width of 10 μs.

►(Section 2.5.1)

$$R_{\text{MIN}} = 0.5 \text{ PW} \times \text{c} = 0.5 \times 10^{-5} \text{ sec} \times 3 \times 10^{8} \text{ m/s} = 1{,}500\text{m} = 1.5 \text{ km}$$

Problem 102–3: Calculate the received signal energy for a radar with 1-kw peak power, a 10% duty cycle, 30-dB peak main beam gain, operating at 5 GHz, with a beamwidth of 2° in a 5-second circular scan, against a 1-m^2 target at 50 km.

►(Section 3.2)

The average power is 100 kw × 0.1 = 10 kw.

The wavelength (λ) = c/F = 3×10^{8} m/s / 5×10^{9} Hz = 0.1m.

The antenna gain (30 dB) = 1,000.

The time on target is equal to the time the target is in the radar beam.

The target is in the beam for (2°/360°) × 5 seconds = 27.8 ms.

(Note that this assumes that there is not significant range change during the beam illumination time.)

$$\text{SE} = \left[P_{\text{AVE}} G^2 \sigma \lambda^2 T_{\text{OT}}\right] / \left[(4\pi)^3 R^4\right]$$
$$= \left[10 \text{ kw} \times 10^6 \times 1\text{m}^2 \times 0.01\text{m}^2 \times 2.7 \times 10^{-2} \text{ sec}\right] / \left[1.98 \times 10^3 \times 6.25 \times 10^6\right]$$
$$= 2.7 \times 10^6 \ / \ 1.23 \times 10^{10} = 2.2 \times 10^{-4} \text{ watt sec}$$

Problem 102–4: Calculate the power into the receiver of the radar described in Problem 102–3 using the decibel formula.

►(Section 3.2)

The peak power is 1 kw = +60 dBm

$$P_R = -103 + P_T + 2\text{G} - 20 \log_{10}(F) - 40 \log_{10}(d) + 10 \log_{10}(\sigma)$$
$$= -103 + 80 + 60 - 20 \log(5{,}000) - 40 \log(50) + 10 \log(1)$$
$$= -103 + 80 + 60 - 74 - 68 + 0 = -105 \text{ dBm}$$

Problem 102–5: Calculate the power leaving the target versus the power arriving at the target for the radar described in Problem 102–3.

►(Section 3.2.1)

Power leaving target /power arriving at target is in effect a gain (G)

$$G = -39 + 20 \log(F) + 10 \log(\text{RCS}) = -39 + 20 \log(5{,}000) + 10 \log(1)$$
$$= -39 + 74 + 0 = 35 \text{ dB}$$

Note that the arriving and leaving signal strengths are normalized to an ideal isotropic antenna at the skin of the target—the signal did not really get bigger by reflecting from the target.

Problem 102–6: Determine the detection range for the radar and target described in Problem 102–3. Assume that the radar receiver sensitivity (including processing gain) is –100 dBm.

►(Section 3.2.2)

Set the received power equal to the sensitivity and solve for 40 log (d)

$$40 \log(d) = -103 + P_T + 2G - 20 \log(F) + 10 \log(\sigma) - \text{Sens}$$
$$= -103 + 80 + 60 - 74 + 0 - (-100) = 63$$

$$d = 10^{[40 \log(d)/40]} = \text{antilog}(63/40) = \text{antilog}(1.575) = 37.6 \text{ km}$$

Problem 102–7: Calculate the sensitivity of an RWR with 20-MHz bandwidth, 10-dB noise figure, and 13-dB required signal-to-noise ratio.

►(Section 3.3.1)

$$\text{kTB} = -114 + 10 \log(20 \text{ MHz} / 1 \text{ MHz}) = -114 + 13 = -101$$

$$\text{Sensitivity} = \text{kTB} + \text{NF} + \text{SNR} = -101 + 10 + 13 = -78 \text{ dBm}$$

Problem 102–8: Determine the range at which an RWR with sensitivity of –45 dBm and an antenna gain of –3 dBi can detect the radar described in Problem 102–3 (at the peak of its main beam).

►(Section 3.3.3)

$$20\log(d) = P_T + G_M - 32 - 20 \log(F) + G_R - \text{Sens}$$
$$= 80 + 30 - 32 - 74 + 3 - (-45) = 52$$

$$d = 10^{[20 \log(d)]/20} = \text{antilog}(52/20) = \text{antilog}(2.6) = 398 \text{ km}$$

Problem 102–9: Determine the range at which an ELINT receiver with a –10-dBi antenna, 10-MHz bandwidth, 3-dB NF, and 13-dB required SNR can detect the radar in Problem 102–3 in a 0-dB sidelobe.

►(Section 3.3.3)

$$\text{kTB} = -114 + 10 \log(10) - 104$$

$$\text{Sens} = \text{kTB} + \text{NF} + \text{SNR} = -104 + 3 + 13 = -88 \text{ dBm}$$

$$20\log(d) = P_T + G_{SL} - 32 - 20\log(F) + G_R - \text{Sens}$$
$$= 80 + 0 - 32 - 74 - 10 - (-88) = 52$$

$$d = 10^{[20\log(d)]/20} = \text{antilog}(52/20) = \text{antilog}(2.6) = 398 \text{ km}$$

Problem 102–10: Determine the Doppler shift that will be seen by a fixed radar at 10 GHz if a target is directly approaching it at 200 m/s.

►(Section 3.6.1)

$$\Delta F = 2(V/c)F = 2(200/3\times10^8)10^{10} = 13.333 \text{ kHz}$$

Problem 102–11: Calculate the Earth surface distance to a target emitter from a single site locator when the elevation angle is 35°, the height of the ionosphere is 100 km.

►(Section 5.2.4)

$$35° = 0.6109 \text{ radians}$$

The radius of the Earth is 6,371 km.

$$d = 2R(\pi/2 - B_R - \sin^{-1}([R\cos B_R/\{R+H\}])$$
$$= 26,(6,271)[\pi/2 - 0.6109 - \arcsin[6,271\cos(35°)/(6,271+100)]$$
$$= 12,542(0.9599 - \arcsin[0.8063] = 12,542[0.9599 - 0.9379] = 276 \text{ km}$$

Problem 102–12: Calculate the line-of-sight spreading loss for a 2-GHz signal at 15 km.

►(Section 5.3.2)

$$L_S = 32 + 20\log(F) + 20\log(d)$$
$$= 32 + 20\log(2{,}000) + 20\log(15) = 32 + 66 + 24 = 122 \text{ dB}$$

Problem 102–13: Calculate the Fresnel zone distance for a 120-MHz signal transmitted from a 2-m-high antenna and received by a 200-m-high antenna.

►(Section 5.3.3)

$$FZ(\text{km}) = (h_T \times h_R \times F) / 24{,}000 = (2 \times 200 \times 120) / 24{,}000$$
$$= 48{,}000 / 24{,}000 = 2 \text{ km}$$

Problem 102–14: Calculate the spreading loss for the signal in problem 102–13 if the path length is 15 km.

►(Section 5.3.3)

The path is longer than the FZ distance, so the two-ray model is appropriate.

$$L_S = 120 + 40\log(d) - 20\log(h_T) - 20\log(h_R) = 120 + 47 - 6 - 46$$
$$= 115 \text{ dB}$$

Problem 102–15: Calculate the knife-edge diffraction loss for a 150-MHz signal transmitted 10 km from the knife edge and received 50 km from the knife edge. The knife edge is 100m higher than the line-of-sight between the transmitter and the receiver.

►(Section 5.3.4)

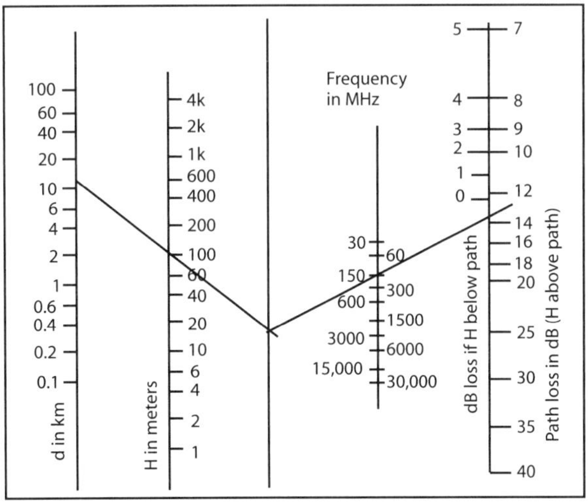

Problem 102–15

Use the nomograph in Figure 5.9 to calculate this loss.

First, calculate the normalized distance "*d*"

$$d = [\text{sqrt}(2) / (1 + d_1 / d_2)] d_1 = 1.414 / (1 + 10 / 50)\ 10\ \text{km}$$
$$= [1.414 / 1.2] \times 10\ \text{km} = 11.8\ \text{km}$$

From the nomograph, the additional knife-edge diffraction loss (above the line-of-sight loss) is 13.5 dB.

Problem 102–16: What is the effective area of an isotropic (i.e., 0-dBi gain) antenna at 250 MHz?

►(Section 5.4)

$$A(\text{dBsm}) = 39 + G - 20 \log(F) = 39 + 0 + 20 \log(250)$$
$$= 39 + 0 - 48 = -9\ \text{dBsm}$$

$$\text{Area}\left(\text{in m}^2\right) = \text{antilog}(-9 / 10) = 0.13\ \text{m}^2$$

Problem 102–17: What is the sensitivity in dBm of a receiver that is specified at 15 μv/m at 500 MHz?

►(Section 5.4)

$$P = -77 + 20 \log(E) - 20 \log(F) = -77 + 20 \log(15) - 20 \log(500)$$
$$= -77 + 24 - 54 = -107\ \text{dBm}$$

Problem 102–18: What is the sensitivity in microvolts per meter for a receiver that is specified at –100 dBm at 150 MHz?

►(Section 5.4)

$$E = \text{antilog}\{[P + 77 + 20 \log(F)] / 20\}$$
$$= \text{antilog}\{[-100 + 77 + 20 \log(150)] / 20\}$$
$$= \text{antilog}\{[-100 + 77 + 44] / 20\} = \text{antilog}\{1.05\} = 11.2\ \mu\text{v/m}$$

Problem 102–19: What is the signal-to-quantization noise for a signal that is digitized with 8 bits?

►(Section 5.6.2)

$$\text{SNR(dB)} = 5 + 3(2m - 1) = 5 + 3(15) = 50 \text{ dB}$$

Problem 102–20: What is the dynamic range (DR) available from an 8-bit digitizer?

►(Section 5.6.2)

$$\text{DR} = 20 \log\left(2^m\right) = 20 \log(256) = 48 \text{ dB}$$

Problem 102–21: Calculate E_b/N_0 for a signal with predetection signal-to-noise ratio of 22 dB, a bit rate of 10,000 bits-per-second, and a bandwidth of 20 kHz.

►(Section 5.6.6)

$$E_b/N_0(\text{dB}) = \text{RFSNR} + 10 \log(\text{bandwidth / bit rate}) = 22 \text{ dB } 10 \log(20\text{k}/10\text{k})$$
$$= 22 + 3 = 25 \text{ dB}$$

Problem 102–22: What J/S is achieved by a 1-kw ERP jammer 50 km from a receiver with a 360° antenna if the desired transmitter has ERP of 10W and is 10 km away? The transmitter, jammer, and receiver are at microwave and all far from the Earth.

►(Section 5.8.1)

The gain of the receiving antenna to the jammer is the same as to the desired transmitter, so the two gain terms go out.

1 kw is + 60 dBm   1W is + 30 dBm

$$\text{J/S} = \text{ERP}_\text{J} - \text{ERP}_\text{S} + 20 \log(d_S) - 20 \log\left(d_J\right) + G_{RJ} - G_R$$

$$= \text{ERP}_\text{J} - \text{ERP}_\text{S} + 20 \log(d_S) - 20 \log\left(d_J\right)$$

$$= 60 - 30 + 20 \log(10) - 20 \log(50) = 60 - 30 + 20 - 34 = 16 \text{ dB}$$

Problem 102–23: A 10-W ERP, 150-MHz transmitter and a receiver are both at 2-m effective antenna height, 10 km apart; both the transmit and receive antennas have 360° coverage. A 1-kw ERP jammer is at 200-m effective antenna height, 50 km from the receiver. What J/S is achieved?

►(Section 5.8.2)

First we must calculate the FZ range for the transmitter and the jammer links.

$$\text{FZ(km)} = (h_T \times h_R \times F)/24{,}000$$

For the transmitter, FZ = (2 × 2 × 150) / 24.000 = 250m

For the jammer, FA = (200 × 2 × 150)/24,000 = 2.5 km

Both lengths require use of the two-ray propagation model

$$J/S = \text{ERP}_J - \text{ERP}_S + 40\log(d_S) - 40\log(d_J) + 20\log(h_J) - 20\log(h_S) + G_{RJ} - G_R$$

Since the receiving antenna has the same gain to the transmitter and jammer, the two gain terms go out.

10W = +40 dBm 1 kw = + 60 dBm

$$J/S = \text{ERP}_J - \text{ERP}_S + 40\log(d_S) - 40\log(d_J) + 20\log(h_J) - 20\log(h_S)$$

$$= 60 - 40 + 40\log(10) - 40\log(50) + 20\log(200) - 20\log(2)$$

$$= 60 - 40 + 40 - 68 + 46 - 6 = 32\text{ dB}$$

Problem 102–24: What is the jamming margin for a direct sequence spread spectrum signal with 1-Mbps chip rate, 1-Kbps bit rate, 0-dB system loss, and a required 15-dB SNR out?

►(Section 5.9)

The code used for spreading is the chip rate, which is 1,000 times the bit rate, so the processing gain is 30 dB. The jamming margin is then:

$$M_J = G_P - L_{SYS} - \text{SNR}_{\text{OUT}} = 30\text{ dB} - 0\text{ dB} - 15\text{ dB} = 15\text{ dB}$$

Problem 102–25: A frequency-hopping transmitter transmits 10W to a 2-dB antenna at a height of 2m. A synchronized receiver with a 2-dB (whip) antenna is 7 km away (and 2m high). The hopping channels are 25-kHz wide and the hopping range is 58 MHz. A variable bandwidth 2-kw noise jammer transmits to a 12-dB log periodic antenna with height of 30m aimed at the receiver from 20-km distance. Two-ray propagation model is appropriate. What jamming bandwidth provides the optimum jamming, and how many received channels are jammed?

►(Section 5.9.1)

$\text{ERP}_J = +63 + 12 = +75$ dBm $\qquad\qquad \text{ERP}_S = +40 + 2 = +42$ dBm

The received power from the desired transmitter is:

$$P_R = \text{ERP}_S - \left[120 + 40\log(d) - 20\log(h_T) - 20\log(h_R)\right] + G_R$$
$$= \text{ERP}_S - 120 - 40\log(d) + 20\log(h_T) + 20\log(h_R) + G_R$$
$$= 42 - 120 - 40 + 6 + 6 + 2 = -104 \text{ dBm}$$

Since frequency hoppers are digital signals, the optimum J/S = 0 dB, so the jammer should put –104 dBm into each receiver channel it is to jam.

The hopping covers 58 MHz in 25-kHz channels, so there are 2,320 hopping channels.

The total jamming power into the receiver is:

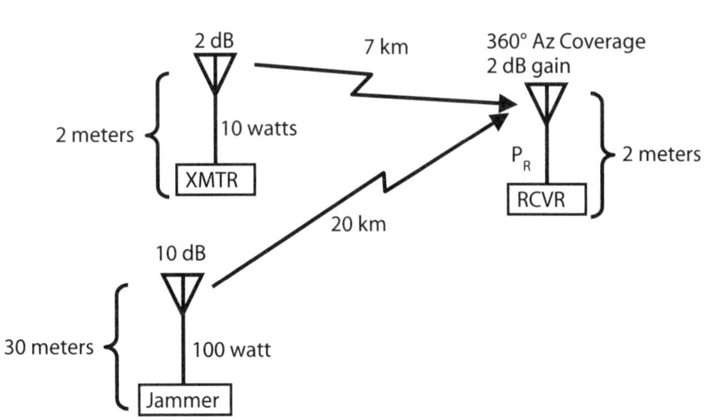

Problem 102–25

$$P_{RJ} = \text{ERP}_\text{J} - \left[120 + 40\log\left(d_J\right) - 20\log\left(h_J\right) - 20\log\left(h_R\right)\right] + G_{RJ}$$
$$= 75 - 120 - 40\log(50) + 20\log(30) + 20\log(2) + 2$$
$$= 75 - 120 - 68 + 30 + 6 + 2 = -75\,\text{dBm}$$

The J/S is –75 – (–104) = 29 dB

29 dB is a factor of antilog(29/10) = 794

The jamming bandwidth would be 794 × 25 kHz = 19.85 MHz.

If the jamming noise is spread over 794 channels, each channel will have 0-dB J/S when the hopping signal happens to hop on that channel.

The ratio of jammed channels is 794/2,320 = 34.2%. Since 33% jamming duty cycle is considered adequate for digital signals, this jammer is effective.

Note that you could also have used the formula:

$$\text{J/S} = \text{ERP}_\text{J} - \text{ERP}_\text{S} + 40\log\left(d_S\right) - 40\log\left(d_J\right) + 20\log\left(h_J\right) - 20\log\left(h_S\right)$$

to determine the J/S directly, then spread the frequency over enough channels to bring it down by 29 dB per channel jammed.

Problem 102–26: If measurements are taken at randomly distributed frequencies over the full frequency range and random azimuths over 360°, find the RMS error, the standard deviation, and the mean error from the following (admittedly small) data set. All data points are in degrees of measurement error from the true angle of arrival.

1.1, –2.0 , +0.5, +0.7, –3.3, –0.2, +1.2, +8, –0.1, + 1.7

►(Section 6.4.1)

First, find the mean error by averaging the error values:

Mean error = sum of errors / 10 = 7.6/10 = +0.76°

Next square each error value and take the average – then the square root of the average.

1.21, 4, 0.25, 0.49, 10.89, 0.04, 1.44, 64, 0.01, 2.89

The sum is 85.22. The mean of the square errors is 8.522. The root mean square (RMS) error is 2.92°.

The standard deviation (the RMS if the mean error is subtracted from each measurement error) is found from

$$\sigma = \text{sqrt}\left[\text{RMS}^2 - \text{mean}^2\right] = 2.92^2 - 0.76^2 = \text{sqrt}[7.95] = 2.81°$$

Problem 102–27: Calculate the spreading loss between a stationery satellite and an Earth station from which the satellite is 5° above the local horizon for a 5-GHz signal.

►(Section 7.4.1)

The path length from the satellite to the above described ground station is 41,408 km.

$$L_S = 32 + 20\log(F) = 20\log(d) = 32 + 20\log(5{,}000) + 20\log(41{,}408)$$
$$= 32 + 74 + 92 = 198 \text{ dB}$$

Problem 102–28: Find the atmospheric loss for a 15-GHz link to a satellite from a ground station that sees the satellite 5° above the horizon.

►(Section 7.4.1)

Using the graph in Figure 7.9, draw a vertical line from 15,000 MHz on the abscissa to the 5° elevation curve. Then draw a horizontal line to the ordinate and read the total atmospheric loss for the link.

3.5 dB

Problem 102–29: Determine the rain loss in a terrestrial link at 10 GHz that passes 25 km through moderate rain.

►(Section 7.6)

Using Table 7.4, determine the correct curve to use in Figure 7.8 (curve C). Then draw a vertical line from 10 GHz on the abscissa of the graph in Figure 7.8 to line C. Then draw a horizontal line to the ordinate and read the rain loss per kilometer.

The rain loss for this link is 0.05 dB/km × 25 km = 1.25 dB.

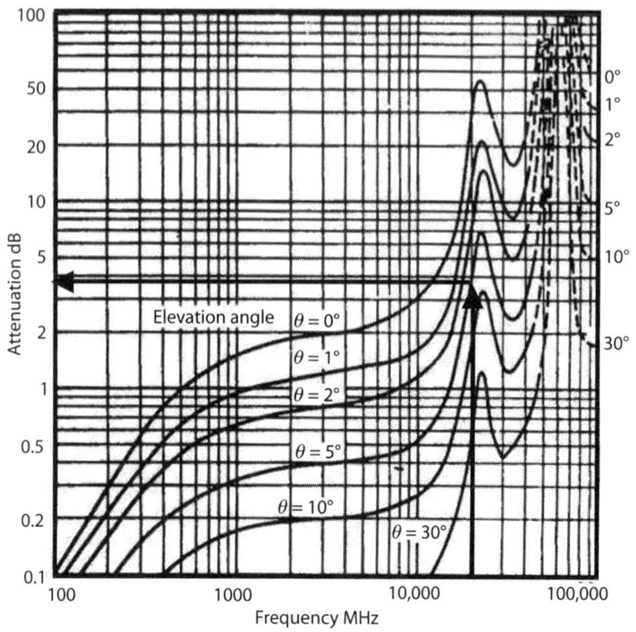

Problem 102–28

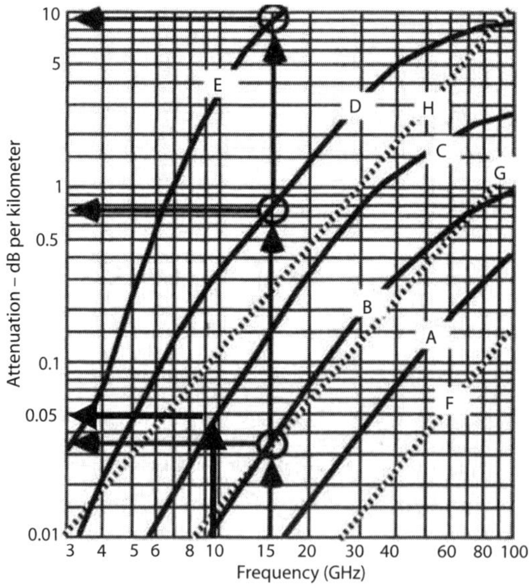

Problem 102–29

Problem 102–30: Determine the rain loss for a 10-GHz satellite to Earth station link with a 5° elevation angle, passing through moderate rain. The 0° isotherm elevation is 3 km.

►(Section 7.3.3)

The path through the rain is from the 0° isotherm to the Earth station, which is

$$d_R = H_{0\ \mathrm{deg}} \,/\, \sin \mathrm{El} = 3\ \mathrm{km}/\sin(5°) = 3\ \mathrm{km}/0.0872 = 34.4\ \mathrm{km}$$

From Problem 102–29, we know that the rain loss is 0.05 dB/km.
The rain loss is 0.05 dB/km × 34.4 km = 1.7 dB.

Problem 102–31: Determine the Earth-station system noise temperature for a receiving system with a 30° antenna elevation, 10-GHz operating frequency, line loss of 6 dB, and receiver noise figure of 2 dB. The ambient temperature is 290K.

►(Section 7.2)

Determine the antenna temperature from the graph in Figure 7.2. Draw a vertical line from 10 GHz on the abscissa to the 30° elevation curve, then left to the ordinate to read the antenna temperature (8K).

The line temperature is: $T_{\mathrm{LINE}} = \left[10^{(L/10)} - 1\right] T_M = \left[10^{6/10} - 1\right] 290\mathrm{K}$
$= \left[\mathrm{antilog}(6/10) - 1\right] \times 290\mathrm{K} = [4 - 1] \times 290\mathrm{K} = 870\mathrm{K}$

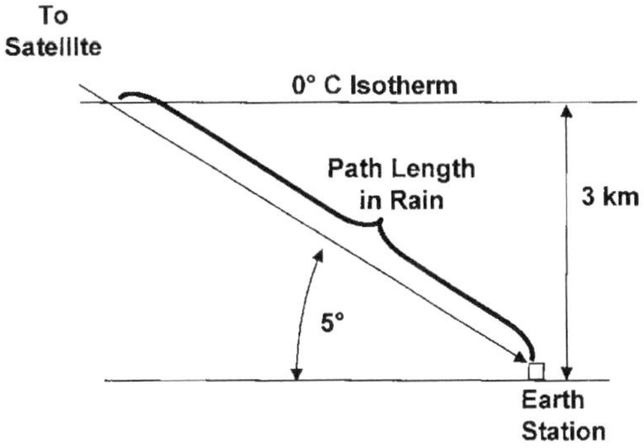

Problem 102–30

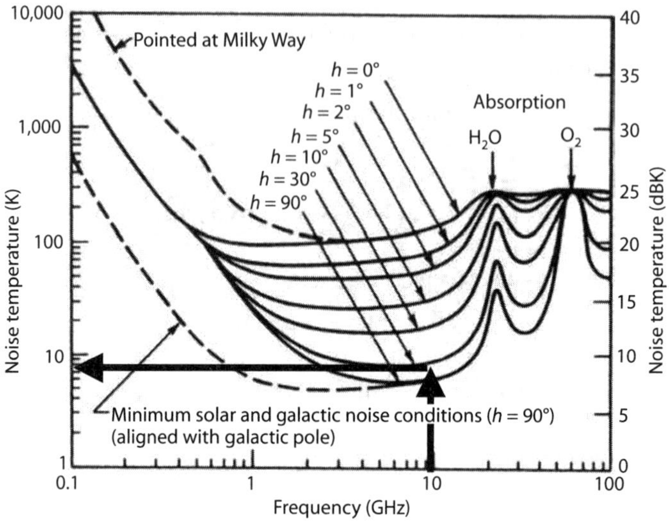

Problem 102–31

The receiver noise temperature is $T_R\left[10^{(\mathrm{NF}/10)} - 1\right]$
$= 290\left[\text{antilog}(\mathrm{NF}/10) - 1\right] = 290\left[\text{antilog}(2/10) - 1\right]$
$= 290\left[1.58 - 1\right] = 170\text{K}$

$$T_S = T_{\mathrm{ANT}} + T_{\mathrm{LINE}} + T_{\mathrm{RX}} = 8\text{K} + 870\text{K} + 170\text{K} = 1{,}048\text{K}$$

Problem 102–32: Calculate C/T in the satellite if the uplink EIRP = 55 dBW, the link loss (spreading + atmospheric + rain) is 200 dB, the receiving antenna gain is 45 dB, and the system noise temperature for the satellite link receiver is 900K.

▸(Section 7.5.1)

To find G/T, convert the antenna gain to its nondecibel form, then divide by the system temperature, then convert back to decibels.

45 dB is $\text{antilog}(45/10) = 31{,}622$
$G/T_S = 31{,}622/900 = 35.1$ which is 15.5 dBi/K
$C/T = \text{EIRP} - L + G/T_S = 55\text{ dBW} - 200\text{ dB} + 15.5\text{ dBi/K}$
$= 55 - 200 + 15.5 = -129.5\text{ dBW/K}$

Problem 102–33: Calculate C/T at the ground station if the downlink EIRP is 60 dBW, the link loss is 204 dB, the receiving antenna gain at the Earth station is 30 dB, and the system noise temperature at the Earth station is 600K.

►(Section 7.5.1)

$$G/T_S = \text{antilog}\left(30 / 10\right) / 600 = 1.7 \text{ dBi/K}$$

$$C/T = \text{EIRP} - L + G/T_S = 60 \text{ dBW} - 204 \text{ dB} + 1.7 = -142 \text{ dBW/K}$$

Problem 102–34: What is the combined signal-to-noise ratio for the uplink and downlink (Problems 102–32 and 102–33) if the bandwidth of the data transferred is 5 MHz?

►(Section 7.5.1)

The inverse of the combined C/T is the sum of the inverses of the uplink and downlink C/T values—which must be converted back to linear form before they can be added.

$C/T \text{ up} = -140 \text{ dBW/K} = \text{antilog}\left(-130 / 10\right) = 1 \times 10^{-13}$

$C/T \text{ down} = -91 \text{ dBW/K} = \text{antilog}\left(-142 / 10\right) = 6.3 \times 10^{-15}$

$1/C/T \text{ combined} = 1/CTup + 1/CTdown = 10^{13} + 1.6 \times 10^{14} = 1.685 \times 10^{14}$

$C/T \text{ combined} = 5.9 \times 10^{-15} = -142.2 \text{ dBW/K}$

$C/kT = C/T - kT = -142.2 + 228.6 = 86.4 \text{ dBHz}$

$C/N = C/kT - 10 \log\left(\text{BW in hertz}\right) = 86.4 - 10 \log(5 \times 10^{6}) = 86.4 - 67 = 19.4 \text{ dB}$

Appendix B: Cross-References to *EW 101* Columns in the *Journal of Electronic Defense*

EW 101

Chapter 1: Introduction
No columns included.

Chapter 2: Basic Mathematical Concepts
July and September 1995 and March and April 2000 columns.

Chapter 3: Antennas
September, October, November, and December 1997 columns.

Chapter 4: Receivers
August, October, and December 1995 and January and April 1996 columns.

Chapter 5: EW Processing
October, November, and December 1998 and January, Feburary, and March 1999 columns.

Chapter 6: Search
January, February, March, April, and part of May 1998 columns.

Chapter 7: LPI Signals
Part of May 1998 column and June, July, August, and September 1998 columns.

Chapter 8: Emitter Location
October 1994 and January, February, March, April, May, and June 1995 columns.

Chapter 9: Jamming
May, June, July, August, November, and December 1996 and January, February, March, and April 1997 columns.

Chapter 10: Decoys
May, June, July, and August 1997 columns.

Chapter 11: Simulation
April, May, June, July, August, September, October, November, and December 1999 and January and February 2000 columns.

EW 102

Chapter 1: Introduction
No columns included.

Chapter 2: Threats
October, November, and December 2001 and January and February 2002 columns.

Chapter 3: Radar Characteristics
May, June, July, August, September, October, November, and December 2000 and January and February 2001 columns.

Chapter 4: Infrared and Electro-Optical Considerations in Electronic Warfare
March, April, May, June, July, August, and September 2001 columns.

Chapter 5: EW Against Communications Signals
June, July, August, September, October, November, and December 2003 and January, February, March, and April 2004 columns.

Chapter 6: Accuracy of Emitter Location Systems
September, October, November, and December 2002 and January, February, March, April, and May 2003 columns.

Chapter 7: Communication Satellite Links
March, April, May, June, and July 2002 columns.

Appendix C: Selected Bibliography

This is a list of some books on electronic warfare and related fields that you might find useful additions to your library. All of these books are on my bookshelf and most of them are used quite frequently—usually to find some particular gem of information.

Adamy, D., *EW 101,* Norwood, MA: Artech House, 2001, ISBN 1-58053-169-5.

Covers the RF aspects of the electronic warfare field using little math. Based on the *EW 101* columns in the *Journal of Electronic Defense.*

Adamy, D., *Introduction to Electronic Warfare Modeling and Simulation,* Norwood, MA: Artech House, 2003, ISBN 1-58053-495-3.

Broad introduction to EW modeling and simulation. Covers terms, concepts, and applications. Includes an introduction to EW sufficient to support the primary material.

Adamy, D., *Practical Communication Theory,* Atwater, CA: Lynx Publishing, 1994, ISBN 1-8885897-04-9.

Describes the one-way communication link and gives simple decibel formulas for working practical intercept problems.

Boyd, J., et al., (eds.), *Electronic Countermeasures*, Palo Alto, CA: Peninsula Publishing, 1961, ISBN 0-932146-00-7.

A compilation of in-depth technical papers on all aspects of EW by recognized experts. Originally a secret textbook prepared under a U.S. Army Signal Corps contract. Declassified in 1973.

Dillard, R., and G. Dillard, *Detectability of Spread Spectrum Signals*, Norwood, MA: Artech House, 1989, ISBN 0-89006-299-4.

Thorough coverage of energy detection approaches to the detection of spread spectrum signals.

Dixon, R., *Spread Spectrum Systems with Commercial Applications*, New York: John Wiley & Sons, 1994, ISBN 0-471-59342-7.

Overviews and mathematical characterizations of spread spectrum signals.

Fahlstron, P., and T. Gleason, *Introduction to UAV Systems*, Columbia, MD: UAV Systems, Inc., 1998, ISBN 995144328.

A very readable coverage of UAV systems: airframe, propulsion, guidance, mission planning, payloads, and data links.

Frater, M., and M. Ryan, *Electronic Warfare for the Digitized Battlefield*, Norwood, MA: Artech House, 2001, ISBN 1-58053-271-3.

Operational focus on the modern electronic battlefield and the appropriate EW techniques. Operational-level descriptions of important new communications EP.

Gibson, J., (ed.), *The Communications Handbook*, Boca Raton, FL: CRC Press, 1977, ISBN 0-8493-8349-8.

Papers on a wide range of communication subjects, including a thorough coverage of propagation models.

Hoisington, D., *Electronic Warfare*, Atwater, CA: Lynx Publishing, 1994, ISBN 1-885897-10-3.

A two-volume text on the whole electronic warfare field using very little math.

Knott, E., J. Shaeffer, and M. Tuley, *Radar Cross Section*, Norwood, MA: Artech House, 1993, ISBN 0-89006-618-3.

Thorough description of radar cross section. Includes the way RCS is modeled and its impact on radar and EW system operation.

Lothes, R., M. Szymanski, and R. Wiley, *Radar Vulnerability to Jamming*, Norwood, MA: Artech House, 1990, ISBN 0-89006-388-5.

Description of important ECM techniques and mathematical description of their effect on radar performance.

Neri, F., *Introduction to Electronic Defense Systems*, Norwood, MA: Artech House, 1991, ISBN 0-89006-553-5.

Nonmathematical coverage of whole EW field. Thorough functional description of threat transmitters.

Pace, P., *Advanced Techniques for Digital Receivers*, Norwood, MA: Artech House, 2000, ISBN 1-58053-053-2.

Graduate-level coverage of digital signals and receivers—design and performance analysis.

Poisel, R., *Introduction to Communication Electronic Warfare Systems*, Norwood, MA: Artech House, 2002, ISBN 1-58053-344-2.

Comprehensive coverage of communication signals and propagation as well as the principles and practice of EW against those signals.

Schleher, D. C., *Electronic Warfare in the Inforamtion Age*, Norwood, MA: Artech House, 1999, ISBN 0-89006-526-8.

Covers the electronic warfare field using both physical and mathematical characterizations. Includes many examples worked in MATLAB 5.1.

Simon, M., et al., (eds.), *Spread Spectrum Communication Handbook*, New York: McGraw-Hill, 1994, ISBN 0-07-057629-7.

Compilation of authoritative papers on spread spectrum communication by experts in the field.

Skolnik, M., *Introduction to Radar Systems*, New York: McGraw-Hill, 2001, ISBN 0072881380.

The authoritative book describing various types of radars and their performance.

Stimson, G., *Introduction to Airborne Radar*, Mendham, NJ: SciTech Publishing, 1998, ISBN 1-901121-01-4.

An extremely gentle yet thorough coverage of radar (not just airborne radar) for those who don't understand (or don't remember) anything about radar.

Vakin, S., L. Shustov, and R. Dunwell, *Fundamentals of Electronic Warfare*, Norwood, MA: Artech House, 2001, ISBN 1-58053-052-4.

Mathematical characterization of EW activities, including thorough coverage of chaff and decoys.

Van Brunt, L., *Applied ECM*, Dunn Loring, VA: EW Engineering, Inc., 1982, ISBN 0-931728-05-3.

A complete and rigorous coverage of electronic countermeasures in three volumes. Available only from the publisher (EW Engineering, Inc., P.O. Box 28, Dunn Loring, VA 22027).

Waltz, E., *Information Warfare Principles and Operation*, Norwood, MA: Artech House, 1998, ISBN 0-89006-511-x.

Comprehensive (nonmathematical) coverage of terms, concepts, and practice of information warfare.

Wiley, R., *Electronic Intelligence: The Interception of Radar Signals*, Dedham, MA: Artech House, 1985, ISBN 0-89006-138-6.

Thorough coverage of the qualitative and quantitative performance of receiving and emitter location systems against a wide range of radar signals.

Wolfe, W., and G. Zissis, (eds.), *The Infrared Handbook*, Washington, D.C.: Office of Naval Research, 1985, ISBN 0-9603590-1-x.

IR theory and practice with lots of tables and graphs characterizing transmission phenomena.

About the Author

David L. Adamy is an internationally recognized expert in electronic warfare, probably mainly because he has been writing the *EW 101* columns for many years. However, in addition to writing the columns, he has been an EW professional (proudly calling himself a "Crow") in and out of uniform for 41 years. As a systems engineer, project leader, program technical director, program manager, and line manager, he has directly participated in EW programs from just above dc to just above light. Those programs have produced systems that were deployed on platforms from submarines to space and met requirements from "quick and dirty" to high reliability.

He holds a B.S. and an M.S. in electrical engineering, both with communication theory majors. In addition to the *EW 101* columns, he has published many technical articles in EW, reconnaissance, and related fields and has 10 books in print (including this one). He teaches electronic warfare-related courses all over the world, and consults for military agencies and EW companies. He is a longtime member of the National Board of Directors of the Association of Old Crows and ran the organization's professional development courses and the technical tracks at its annual technical symposium until he was elected national president in 2001.

He has been married to the same long-suffering wife for 43 years (she deserves a medal for putting up with a classical nerd that long) and has four daughters and eight grandchildren. He claims to be an okay engineer, but one of the world's truly great fly fishermen.

Index

Recent Titles in the Artech House Radar Library

David K. Barton, Series Editor

Advanced Techniques for Digital Receivers, Phillip E. Pace

Airborne Pulsed Doppler Radar, Second Edition, Guy V. Morris and Linda Harkness, editors

Bayesian Multiple Target Tracking, Lawrence D. Stone, Carl A. Barlow, and Thomas L. Corwin

Beyond the Kalman Filter: Particle Filters for Tracking Applications, Branko Ristic, Sanjeev Arulampalam, and Neil Gordon

Computer Simulation of Aerial Target Radar Scattering, Recognition, Detection, and Tracking, Yakov D. Shirman, editor

Design and Analysis of Modern Tracking Systems, Samuel Blackman and Robert Popoli

Detecting and Classifying Low Probability of Intercept Radar, Phillip E. Pace

Digital Techniques for Wideband Receivers, Second Edition, James Tsui

Electronic Intelligence: The Analysis of Radar Signals, Second Edition, Richard G. Wiley

Electronic Warfare in the Information Age, D. Curtis Schleher

EW 101: A First Course in Electronic Warfare, David Adamy

EW 102: A Second Course in Electronic Warfare, David L. Adamy

Fourier Transforms in Radar and Signal Processing, David Brandwood

Fundamentals of Electronic Warfare, Sergei A. Vakin, Lev N. Shustov, and Robert H. Dunwell

Fundamentals of Short-Range FM Radar, Igor V. Komarov and Sergey M. Smolskiy

Handbook of Computer Simulation in Radio Engineering, Communications, and Radar, Sergey A. Leonov and Alexander I. Leonov

High-Resolution Radar, Second Edition, Donald R. Wehner

Introduction to Electronic Defense Systems, Second Edition, Filippo Neri

Introduction to Electronic Warfare, D. Curtis Schleher

Introduction to Electronic Warfare Modeling and Simulation, David L. Adamy

Introduction to RF Equipment and System Design, Pekka Eskelinen

Microwave Radar: Imaging and Advanced Concepts, Roger J. Sullivan

Millimeter-Wave Radar Targets and Clutter, Gennadiy P. Kulemin

Modern Radar System Analysis, David K. Barton

Multitarget-Multisensor Tracking: Applications and Advances Volume III, Yaakov Bar-Shalom and William Dale Blair, editors

Principles of High-Resolution Radar, August W. Rihaczek

Principles of Radar and Sonar Signal Processing, François Le Chevalier

Radar Cross Section, Second Edition, Eugene F. Knott et al.

Radar Evaluation Handbook, David K. Barton et al.

Radar Meteorology, Henri Sauvageot

Radar Reflectivity of Land and Sea, Third Edition, Maurice W. Long

Radar Resolution and Complex-Image Analysis, August W. Rihaczek and Stephen J. Hershkowitz

Radar Signal Processing and Adaptive Systems, Ramon Nitzberg

Radar System Performance Modeling, G. Richard Curry

Radar Technology Encyclopedia, David K. Barton and Sergey A. Leonov, editors

Range-Doppler Radar Imaging and Motion Compensation, Jae Sok Son et al.

Space-Time Adaptive Processing for Radar, J. R. Guerci

Theory and Practice of Radar Target Identification, August W. Rihaczek and Stephen J. Hershkowitz

Time-Frequency Transforms for Radar Imaging and Signal Analysis, Victor C. Chen and Hao Ling

For further information on these and other Artech House titles, including previously considered out-of-print books now available through our In-Print-Forever® (IPF®) program, contact:

Artech House
685 Canton Street
Norwood, MA 02062
Phone: 781-769-9750
Fax: 781-769-6334
e-mail: artech@artechhouse.com

Artech House
46 Gillingham Street
London SW1V 1AH UK
Phone: +44 (0)20 7596-8750
Fax: +44 (0)20 7630-0166
e-mail: artech-uk@artechhouse.com

Find us on the World Wide Web at:
www.artechhouse.com